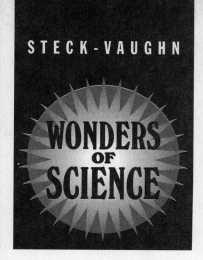

STECK-VAUGHN

WONDERS OF SCIENCE

Joan S. Gottlieb

The Earth and Beyond

TEACHER'S EDITION

STECK-VAUGHN
COMPANY
ELEMENTARY • SECONDARY • ADULT • LIBRARY

C O N T E N T S

Scope and Sequence T3

Wonders of Science Program T4
   About the Program
   Features of the Program
   Using the Program

Correlation to *Benchmarks for Science Literacy* T5

Master Materials List: *Explore & Discover* Activities T6

Teaching Strategies T7

Mastery Test Part A, units 1–4 .... T15

Mastery Test Part B, units 5–8 .... T17

Mastery Test Answer Key T19

Just for Fun Blackline Masters T20

Just for Fun Answer Key T28

Special Projects T30

Related Steck-Vaughn Products T32

Unit 1: The Solar System 4

Unit 2: Weather 26

Unit 3: Land and Water 46

Unit 4: Regions of Earth 68

Unit 5: Shaping the Surface 82

Unit 6: The Changing Earth 96

Unit 7: Materials of Earth 110

Unit 8: Conservation 126

Glossary 138

Wonders of Science Program

■ **The Earth and Beyond**

❏ **The Human Body**

❏ **Land Animals**

❏ **Matter, Motion, and Machines**

❏ **Plant Life**

❏ **Water Life**

About the Author

Joan S. Gottlieb taught at the elementary and secondary levels
for more than ten years. She received a Ph.D. in education
from the University of South Carolina. Dr. Gottlieb holds B.A.,
B.S., and M.Ed. degrees from the University of Minnesota.

ISBN 0-8114-7496-8

Copyright © 1996 Steck-Vaughn Company

4 5 6 7 8 9 10 PO 99 98

Scope and Sequence

	UNIT 1	UNIT 2	UNIT 3	UNIT 4	UNIT 5	UNIT 6	UNIT 7	UNIT 8
The Human Body	How the Body Is Organized	Body Systems	More Body Systems	The Nervous System	Human Reproduction	Diseases	Avoiding Health Problems	First Aid
Land Animals	Animal Adaptations	Invertebrate Animals	Amphibians	Reptiles	Birds	Mammals	Conservation	
Plant Life	Plants Are Living Things	Flowers	Grasses and Cereals	Trees	Plants Used as Food	Plant Adaptations	Plant Products	Conservation
Water Life	Water Environments	Water Plants	Invertebrate Water Animals	Fish	Water Reptiles	Water Birds	Water Mammals	Conservation
The Earth and Beyond	The Solar System *Communicating Interpreting Data Making Models*	Weather *Communicating Classifying Inferring*	Land and Water *Communicating Classifying Making Models*	Regions of Earth *Communicating Making Models Inferring*	Shaping the Surface *Communicating Observing Inferring*	The Changing Earth *Communicating Interpreting Data Classifying*	Materials of Earth *Communicating Classifying Making Models*	Conservation *Communicating Classifying Observing*
Matter, Motion, and Machines	Matter	Changes in Matter	Nature's Energies	Sound and Light	Magnetism and Electricity	Motion and Forces	Machines	Technology

About the Program

Wonders of Science was developed for students with special needs, including students in special education programs, students receiving remedial instruction, and students in regular classes who read at a second-to-third-grade reading level. This six-book program presents current science topics in a simple, straightforward manner.

Wonders of Science was written by a classroom teacher who has faced the difficulties of teaching junior and senior high school special-needs students without suitable teaching materials. Like most teachers of special-needs classes, she invested uncountable hours preparing materials that would capture her students' interest and offer them a chance to succeed.

Wonders of Science is a result of field testing a variety of activities in classes with a wide range of reading abilities, intellectual capacities, ages, maturity levels, social backgrounds, interests, attention spans, and physical capabilities. The end result is a program designed to appeal to a wide variety of special-needs students. This program offers a structured curriculum, yet it helps students develop the much needed skill of working independently.

Features of the Program

❏ All lessons are complete on one or two pages. Brief reading selections provide students with successful experiences. Students with short attention spans are not frustrated by lesson length.

❏ Scientific vocabulary appears in boldfaced type. Each vocabulary word is also defined in the Glossary at the back of the book.

❏ In the Teacher's Edition, words that may be unfamiliar or difficult for students are underlined. These words may need further explanation.

❏ Assessment in the form of exercises is included in every lesson to check retention of important scientific concepts. Unit and Mastery Tests are also provided.

❏ A one-page review at the end of each unit helps prepare students for the unit test. Student performance on the review can be used to plan reteaching activities based on individual needs before students take the unit test.

❏ *Explore & Discover,* a self-directed, hands-on activity at the end of each unit, provides students with an opportunity to develop process skills and reinforce concepts taught in the unit. These activities call for easy-to-obtain and use materials. Students should be more comfortable with familiar materials than with expensive scientific hardware and apparatus. Yet the physical and cognitive skills used are the same.

❏ In each unit test, main ideas and scientific concepts are presented in a format that incorporates standardized test items as well as open-ended sections.

❏ Teaching Strategies for each lesson are provided in the Teacher's Edition. These strategies, in the form of activities, provide support, reinforcement, and extension of concepts through real-life application, hands-on activities, or discussion.

❏ A *Just for Fun* page for each unit is included in the Teacher's Edition. This blackline master presents key scientific concepts from the unit in a puzzle format and can be used as reinforcement.

❏ Special Projects in the Teacher's Edition are motivational activities that supplement and provide enrichment to each unit. Read through the projects before beginning a unit to determine when each is most helpful to your class.

❏ A two-part Mastery Test is included in the Teacher's Edition for assessing student comprehension of key concepts. The Mastery Test is written in the same format as the unit tests to ease test anxiety and promote student success.

Using the Program

The *Wonders of Science* program was designed to be used flexibly. You are encouraged to use the materials in a way that best meets the needs of your students. For example, you may wish to introduce the vocabulary words first, and then have students read each lesson independently. Or, for students who are able to work independently, you may encourage use of the Glossary as an aid for defining new scientific vocabulary.

All exercise directions are shown in color and are, therefore, easily distinguished. Each exercise format is repeated throughout the text so that students become familiar with the directions.

The design of each page is friendly and easy to use. All text is contained in a box to help students focus on the lesson. A color band runs down the side of the reading passage to cue the student on what to read. Directions are set in color to cue the student when a new activity starts. Photographs and art are used to reinforce and illustrate scientific concepts.

Correlation to
Benchmarks for Science Literacy

In an effort to guide reform in science, mathematics, and technology education, the American Association for the Advancement of Science published the report *Benchmarks for Science Literacy* in 1993. This report specifies how students should progress toward science literacy, recommending what they should know and be able to do by the time they reach certain grade levels. Intended as a tool for designing curriculum, *Benchmarks* describes levels of understanding and ability for *all* students. Steck-Vaughn's *Wonders of Science* provides special-needs students with opportunities to build the common core of learning recommended by *Benchmarks*. In *The Earth and Beyond,* students can also explore their individual interests and abilities by examining topics such as earthquakes, natural resources, and patterns of stars.

Wonders of Science: The Earth and Beyond will be useful in helping students attain the following specific science literacy goals set out in *Benchmarks*:

Chapter 3: THE NATURE OF TECHNOLOGY
Section A: Technology and Science
Technology is essential to science for such purposes as access to outer space and other remote locations, sample collection and treatment, measurement, data collection and storage, computation, and communication of information.

Section A: The Universe
Nine planets of very different size, composition, and surface features move around the sun in nearly circular orbits. Some planets have a great variety of moons and even flat rings of rock and ice particles orbiting around them. Some of these planets and moons show evidence of geologic activity. The earth is orbited by one moon, many artificial satellites, and debris.

Section B: The Earth
Some minerals are very rare and some exist in great quantities, but—for practical purposes—the ability to recover them is just as important as their abundance. As minerals are depleted, obtaining them becomes more difficult. Recycling and the development of substitutes can reduce the rate of depletion but may also be costly.

Section B: The Earth
The benefits of the earth's resources—such as fresh water, air, soil, and trees—can be reduced by using them wastefully or by deliberately or inadvertently destroying them. The atmosphere and the oceans have a limited capacity to absorb wastes and recycle materials naturally. Cleaning up polluted air, water, or soil or restoring depleted soil, forests, or fishing grounds can be very difficult and costly.

Section B: The Earth
The earth is mostly rock. Three-fourths of its surface is covered by a relatively thin layer of water (some of it frozen), and the entire planet is surrounded by a relatively thin blanket of air. It is the only body in the solar system that appears able to support life. The other planets have compositions and conditions very different from the earth's.

Section B: The Earth
Because the earth turns daily on an axis that is tilted relative to the plane of the earth's yearly orbit around the sun, sunlight falls more intensely on different parts of the earth during the year. The difference in heating of the earth's surface produces the planet's seasons and weather patterns.

Section B: The Earth
The moon's orbit around the earth once in about 28 days changes what part of the moon is lighted by the sun and how much of that part can be seen from the earth—the phases of the moon.

Section B: The Earth
Fresh water, limited in supply, is essential for life and also for most industrial processes. Rivers, lakes, and groundwater can be depleted or polluted, becoming unavailable or unsuitable for life.

Section C: Processes That Shape the Earth
The interior of the earth is hot. Heat flow and movement of material within the earth cause earthquakes and volcanic eruptions and create mountains and ocean basins. Gas and dust from large volcanoes can change the atmosphere.

Section C: Processes That Shape the Earth
Some changes in the earth's surface are abrupt (such as earthquakes and volcanic eruptions) while other changes happen very slowly (such as uplift and wearing down of mountains). The earth's surface is shaped in part by the motion of water and wind over very long times, which act to level mountain ranges.

Section C: Processes That Shape the Earth
Sediments of sand and smaller particles (sometimes containing the remains of organisms) are gradually buried and are cemented together by dissolved minerals to form solid rock again.

Section C: Processes That Shape the Earth
Sedimentary rock buried deep enough may be reformed by pressure and heat, perhaps melting and recrystallizing into different kinds of rock. These re-formed rock layers may be forced up again to become land surface and even mountains. Subsequently, this new rock too will erode. Rock bears evidence of the minerals, temperatures, and forces that created it.

Section C: Processes That Shape the Earth
Human activities, such as reducing the amount of forest cover, increasing the amount and variety of chemicals released into the atmosphere, and intensive farming, have changed the earth's land, oceans, and atmosphere. Some of these changes have decreased the capacity of the environment to support some life forms.

Chapter 10: HISTORICAL PERSPECTIVES
Section A: Displacing the Earth from the Center of the Universe
Telescopes reveal that there are many more stars in the night sky than are evident to the unaided eye, the surface of the moon has many craters and mountains, the sun has dark spots, and Jupiter and some other planets have their own moons.

Chapter 12: HABITS OF MIND
Section D: Communication Skills
Organize information in simple tables and graphs and identify relationships they reveal.

Section D: Communication Skills
Read simple tables and graphs produced by others and describe in words what they show.

Section D: Communication Skills
Locate information in reference books, back issues of newspapers and magazines, compact disks, and computer databases.

Master Materials List
Explore & Discover Activities

The following list is designed to help you acquire and keep track of materials for the **Explore & Discover** activities in this book. These materials are commonly available in classrooms, or may be purchased at supermarkets or discount stores, or brought from students' homes.

The quantities and groupings may be adjusted depending on the needs of your class, and in some cases you may want to substitute materials. The page numbers of the activities are provided to help your planning.

	Item	Quantity	Notes
Unit 1 **The Solar System** *page 24*	clay, modeling toothpicks balls, tennis	1/4 stick per pair 1 per pair 1 per pair	Also used in Units 3, 4, 7
Unit 2 **Weather** *page 44*	thermometers paper, unlined	4 per pair 1 sheet per pair	See note below*
Unit 3 **Land and Water** *page 66*	containers, plastic, 1 gal. clay, modeling water paper, unlined toothpicks tape, clear scissors	1 per student 1 stick per student 1/2 cup per student 1 sheet per student 6 per student several rolls per class 1 pair per student	Also used in Unit 4 Also used in Units 1, 4, 7
Unit 4 **Regions of Earth** *page 80*	container, plastic, 1 gal. clay, modeling paper, unlined, or tracing paper	1 per pair 1 stick per pair 1 sheet per pair	Also used in Unit 3 Also used in Units 1, 3, 7
Unit 5 **Shaping the Surface** *page 94*	sand, damp pans, pie, unbreakable cartons, milk, small water buckets	1 cup per pair 1 per pair 1 per pair enough to fill cartons 1 per pair	
Unit 6 **The Changing Earth** *page 108*	felt squares, red, yellow, and blue pencils, colored, red, yellow, and blue paper, unlined	2 each per student 1 each per student 1 sheet per student	
Unit 7 **Materials of Earth** *page 124*	clay, modeling shells, small stones, small leaves	1/2 stick per student 1 per student 1 per student 1 per student	Also used in Units 1, 3, 4
Unit 8 **Conservation** *page 136*	paper, construction, 11" x 18" markers, various colors	1 sheet per lab team 1 set per lab team	

*Calibrating Thermometers
Before beginning the activity, put all the thermometers in the same place and leave them for about 10 minutes. If the temperature readings vary, choose one thermometer and put a sticker on it that says "+/- 0" to indicate it is accurate. Then put stickers on the others that say +/- the number of degrees, Fahrenheit or Celsius, it would take to reach the accurate temperature.

UNIT 1 — The Solar System

Unit 1 describes the objects that make up our solar system. The sun and other stars are discussed. The difference between stars and planets is pointed out. Characteristics of the inner planets and outer planets are given. Earth is discussed in relation to the other planets. Its year, day and night, and seasons are explained in terms of orbit rotation, revolution, and tilt. Finally, the unit explains how people are exploring the solar system and beyond.

Page 4—The Parts of the Solar System
Materials: paper and marking pens, encyclopedia or book on space

Have students refer to an encyclopedia or a book on space for models of the solar system. Then have them draw a model of their own. They should include the sun, the nine planets, Earth's moon, a comet, and asteroids. The students should label each part of their solar system model. (At this time, students should not be concerned with drawing the planets to scale or representing their distances from the sun to the correct scale. However, the order of the planets should be correct.)

Page 6—The Sun and Other Stars

Explain to students that light travels faster than anything else in the known universe—186,000 miles every second. Explain that the stars are so far away from Earth that astronomers use a unit called the light-year to measure the distance. One light-year is the distance light travels in 1 year (almost 6 trillion miles). After the sun, the nearest star to Earth is Proxima Centauri. It is over 4 light-years away. This means the light from Proxima Centauri, traveling 186,000 miles each second, takes 4 years to reach Earth.

Page 8—Patterns of Stars
Materials: flashlight, shoe box, tracing paper, tape, construction paper, scissors, book on stars, straight pin

Help students make a constellation projector. First, select several constellations from a book on the stars. Then trace these constellations onto tracing paper. Tape the tracings onto a sheet of construction paper. Using a straight pin, poke small holes through the two pieces of paper where each star is marked. Cut out one end of a shoe box and tape the paper over the cutout end. Place a flashlight in the box facing the cutout end. Turn the flashlight on and put the lid on the box. Turn down the lights in the room and aim the box at a blank wall to see the constellations.

Page 10—The Inner Planets

Divide the class into two teams for a Planet Quiz. Ask questions about the planets described in the lesson and have the students respond as a team. The team that answers more questions correctly is the winner.

Page 12—Earth

Students should understand that, as far as we know, Earth is the only planet that has living things. Plants and animals can live on Earth because the temperature is neither too hot nor too cold. Living things need the water of Earth and the gases of the atmosphere. Other planets have atmospheres, but they are made up of gases that do not support living things.

Page 14—Earth's Moon
Materials: lamp, ball

Students can observe the way sunlight produces the phases of the moon. Tell students that the lamp represents the sun, the ball is the moon, and that a student will be Earth. Have a student volunteer sit with the lamp at eye level, about 2 feet away. Hold the ball so that it is equidistant from the lamp and the student's eyes. Have the student describe the amount of ball that is lit. Then put the lamp at the student's side. Hold the ball 1 foot in front of the student's eyes. Have the student describe the amount of ball that is lit now. Finally, put the lamp behind the student. Hold the ball to one side of the student's face. The student should describe the lighted part of the ball once again.

Page 16—The Outer Planets
Materials: thumbtack, string, cardboard, ruler, scissors

To appreciate the difference in size among the planets, students can make cutouts of each planet. Have them tie a piece of string to the end of a pencil. Then have them measure the following lengths on the string, starting from the pencil.

Mercury—$\frac{1}{2}$ in. Mars—$\frac{3}{4}$ in. Uranus—$4\frac{1}{2}$ in.

Venus—$1\frac{1}{2}$ in. Jupiter—$13\frac{1}{2}$ in. Neptune—$4\frac{1}{2}$ in.

Earth—$1\frac{1}{2}$ in. Saturn—11 in. Pluto—$\frac{3}{4}$ in.

Have them place a thumbtack through the string and into a sheet of cardboard at the first length and trace a circle for the planet. Repeat this procedure on separate sheets of cardboard for the remaining planets. Then have students cut out the circles and label them with the names of the planets.

Page 18—Exploring Space

Materials: newspapers or magazines, notebook

Students may want to start a Space Notebook. In the notebook, they can keep articles on space exploration from newspapers and magazines. Students may contact an office of NASA for additional information on recent discoveries.

Page 20—A Year

Remind students that we make adjustments in time, such as the change called daylight saving time. In spring, clocks are set forward 1 hour to take advantage of the extra hours of daylight. In the fall, clocks are set back 1 hour as daylight hours decrease.

Page 21—Day and Night

Materials: globe, flashlight

Use a globe to show students how the rotation of Earth causes day and night. Point out the equator and the axis. Then point out where you live. Darken the room and shine a flashlight on the globe. As the globe turns, have students decide whether it is day or night where they live.

Page 22—The Seasons

Materials: globe

Show students the portion of the globe that defines the Northern Hemisphere. Point out that the United States and Canada are in the Northern Hemisphere. Tilt the globe so that the Northern Hemisphere is pointing toward the class. Explain that the day that corresponds with the greatest tilt toward the sun is the longest day of the year. This occurs on June 20 or 21. Then tilt the globe so that the Northern Hemisphere is pointing away from the class. Explain that the day that corresponds with the greatest tilt away from the sun is the shortest day of the year. This occurs on December 21 or 22. Orient the globe to show times when day and night are the same length. At these times, neither the North Pole nor the South Pole is tilted toward the sun.

Page 24—Model the Seasons

This activity has students implement the process skill *Making Models* as they show the movement of Earth around the sun and demonstrate the change of seasons. Providing a class demonstration with the students acting out the roles may be helpful.

UNIT 2 Weather

Unit 2 describes the conditions that produce weather on Earth. The layers of the atmosphere are described. The movement of water through the water cycle is described, and the formation of clouds and precipitation is explained. Changes in the weather—both normal and dramatic—are explained in terms of changing air pressure and the movement of air masses. Finally, the methods and tools used by meteorologists are outlined.

Page 26—Weather and the Atmosphere

Let students know that a large hole in the ozone layer has been found to occur every year over Antarctica. Some scientists believe that the damage to the ozone layer is caused by the release of chemicals called chlorofluorocarbons. These chemicals are used in spray cans, refrigerators, and manufacturing processes. Some countries, including the United States, now ban the use of these chemicals in spray cans.

Page 28—The Water Cycle

Materials: glass jar with airtight lid, lamp, ice

Have students make a model of the water cycle. Have them place a few drops of water in a glass jar and then seal the jar. Put the jar under a lamp. When the drops have evaporated and changed to water vapor, have students place the jar in a bowl of ice. After a few minutes, have them wipe off the outside of the jar. They should notice that the vapor inside the jar has condensed back into drops.

Page 30—Clouds

Materials: glass jar with lid, ice, flashlight

Have students make a cloud. Have them fill a jar halfway with hot water and then seal the jar with its lid. Then have them place ice on the lid. Dim the lights in the room. Shine a flashlight into the jar. Ask students why a cloud is forming in the jar. (The air in the jar is warmed by the hot water. The warm air rises and is cooled by the ice. Cool air cannot hold as much water as warm air, so droplets of water condense, forming a cloud.)

Page 32—Precipitation

Materials: almanac

Have students use an almanac to research the normal amount of precipitation that falls in their area each month. Then have them find the same information for another part of the country. Help students make a chart of their findings. You may also want to prepare a bar graph on the chalkboard using the students' data. Then help students read the information in the graph.

Page 34—Humidity

Ask students when clothes hung outdoors will dry more quickly—on a summer day when the humidity is high or on a day when the humidity is low. (Students should understand that when the humidity is high, the air is holding a great deal of water. So the water in the wet clothes will not evaporate into the air easily. It will evaporate more quickly on a day with lower humidity.)

Page 35—Temperature

Materials: Fahrenheit and Celsius thermometers, glass, water

Help students practice reading thermometers. Place a Fahrenheit and a Celsius thermometer in a glass of warm water. Have a student read the temperature on the Fahrenheit scale. Have another student read the temperature on the Celsius scale. Then repeat the activity with cold water.

Page 36—What Makes the Weather Change?

Precipitation along a cold front is usually heavy. The air rises quickly and the water vapor condenses quickly, resulting in heavy precipitation. Precipitation along a warm front is usually lighter, but lasts for a longer time.

Page 38—Air Pressure

Materials: two balloons, two small glass bottles, two pie pans, ice

Stretch a balloon over the mouth of each bottle. Put one bottle in a pie pan filled with ice. Put the second bottle in a pie pan and place it on a heat source or near a sunny window. Have students note what happens to the balloons as the air in one bottle is cooled while the air in the other bottle is warmed. Ask students where the air pressure is greater. (As air is heated, its particles spread out and so the pressure is lower. In cold air, the particles are close together and the pressure is higher.)

Page 40—Storms

Materials: newspapers

Students can research recent storms that have been reported in the newspaper. Have them cut out the stories and make a bulletin board display. Have them post the stories under the headings *Thunderstorm, Tornado,* or *Hurricane.*

Page 42—Tracking the Weather

Interested students can give their own weather reports to the class. Have them listen to a television or radio weather report the night before they plan to report on the weather. Then have them cut out a large weather map from a morning newspaper. If possible, reproduce the map and give a copy to each student.

Page 44—Record Temperatures

This activity reinforces the process skill of *Inferring* as students measure the temperature of the classroom in several places to discover that the readings vary between high, low, dark, and light places. Students may also wish to record the temperature outside for an additional comparison.

UNIT 3 Land and Water

Unit 3 looks at some of the areas of land and water that make up Earth's surface. Bodies of fresh water, including rivers, lakes, and ponds, are described. Wells are discussed as ways to reach underground water. Oceans, bodies of salt water, are discussed. The features of the ocean floor are described, as well as kinds of life that can be found in the oceans. The major continents are discussed and features, such as mountains, plains, and plateaus, are described.

Page 46—Rivers

Materials: encyclopedia, almanac

Students can research a major river for a report to the class. Possible subjects include the Mississippi, the Missouri, the Colorado, the Rio Grande, the Yukon, the Hudson, the Nile, the Amazon, the Yangtze, and the Volga. Students should learn where the river begins, its length, and where its mouth is located. Students can also include any interesting historical facts they discover.

Page 48—Lakes and Ponds

Tell students that other bodies of fresh water include marshes, swamps, and bogs. A marsh is a large, shallow body of water that usually has a dense cover of reeds, sedges, and cattails. A swamp is shallow like a marsh, but instead of reeds the dominant plants are trees, such as mangrove or cypress. In a bog, the major plants are usually sphagnum mosses. Bog water is very acidic and low in nutrients. Some bog plants, like the Venus's-flytrap, sundew, and pitcher plant, supplement their nutrients by trapping insects. Marshes and bogs are often formed when ponds fill in with silt and debris. Swamps are often caused by the overflow from rivers.

Page 50—Wells

Materials: small plastic cup, funnel, clay, rubber tubing, large plastic container, pencil

Have students make a model of a flowing artesian well. Add about an inch of water to a small plastic cup. Place one end of the rubber tubing in the cup. Fill the rest of the cup with clay, molding it around the tubing and the edges of the cup. Have one student hold the cup over a plastic container. Then, using a funnel, carefully pour water into the tubing until it is filled. Have another student gently dig down into the clay with a pencil until the "water table" is reached. The artesian well should begin to flow.

Page 52—Caves

Caves may appear to have little to offer animal life, but many animals can be found there. Some animals, such as bats, mountain lions, and peccaries, hide in caves during the day. Other animals live in caves their entire lives. Some kinds of salamanders, fish, crickets, and crayfish never leave their cave homes. The blind cave fish even loses its ability to see, since eyes are of little help in the darkness of a cave.

Page 54—The Oceans

Materials: world map from an atlas or encyclopedia, graph paper (with large squares), marking pens, scissors

Help students visualize how much of Earth is covered by water. Have students trace a world map onto a sheet of graph paper. Have them color the land areas one color and the oceans another. Then have them cut out the land areas and separate them from the oceans. Help students count the number of complete squares in the land areas and the number in the oceans. The number of ocean squares should be between three and four times the number of land squares.

Page 56—Along the Ocean Floor

Tell students that the deepest place in the world is the Mariana Trench in the Pacific Ocean. This trench is more than 35,000 feet deep in certain areas. By contrast, the tallest mountains in the world are about 29,000 feet high.

Page 58—Life in the Oceans

Materials: encyclopedia or book on the oceans, poster paper, marking pens

Have students research the kinds of plants and animals found in the ocean. Using drawings of these living things, have students make a bulletin board display that shows in which level they are found.

Page 60—Continents

Materials: encyclopedia

Have small groups of students research each of the continents. Students should list facts about each continent, such as the area and population of the continent, the tallest mountain, the longest river, and any interesting animals found there. The information can be made into a chart and posted on the bulletin board.

Page 62—Mountains

Materials: encyclopedia

Divide the class into two groups. Have one group research the Rocky Mountains. Have the second group research the Appalachian Mountains. Each group should learn how long the mountain range is, how high the highest point is, what plants and animals live in the mountains, and any interesting historical facts they can find.

Page 64—Plains and Plateaus

Materials: encyclopedia or geography book

Have students locate the Great Plains on a map of North America. The Great Plains extend from northern Canada to southern Texas. They are about 400 miles east of the Rocky Mountains. Deposits of coal, oil, and natural gas are found beneath the Great Plains.

Page 66—Make a Model Island

This activity will encourage students to use the process skill *Making Models* as they replicate a model of the ocean floor. Students may wish to further investigate islands by reading library books or encyclopedias.

UNIT 4 Regions of Earth

Unit 4 outlines the characteristics of the three major climate zones. It explains how factors such as altitude, latitude, and nearness to ocean currents affect climate. It describes different environments within these zones: deserts, forests, grasslands, mountains, and tundra.

Page 68—Climate Zones
Materials: globe

Have students look at a globe and locate the equator. Point out the lines of latitude north and south of the equator. The equator is 0° latitude. The North Pole is 90° north latitude and the South Pole is 90° south latitude. The other latitude lines fall between these two points. Ask students to locate the latitude line closest to where they live.

Page 70—Deserts
Materials: encyclopedia

Have students research one of the great deserts of the world. These include the Sahara, Mohave, Kalahari, Atacama, Gobi, and Great Victoria Desert. Students can prepare a report that describes the location and size of the desert, the yearly rainfall, and any unusual plants and animals.

Page 72—Forests
Materials: globe or map of South America

Have students use a map to find the Amazon River in South America. Tell them that the tropical rain forest around the Amazon River is about two thirds the size of the United States. Between 50 and 120 inches of rain fall in a year. Scientists estimate that the rain forest contains over 50,000 species of plants. Some of the world's largest reptiles and snakes also live in the forest.

Page 74—Grasslands

Tell students that over 50 different species of hooved mammals live on the African grasslands. The reason so many can live in the same place is because each lives in a different way. Some, like the giraffe, browse on bushes and small trees. Others, like zebra and African buffalo, eat tall grasses. Certain antelope, like gazelle and impala, eat only the fresh tips of young grasses and herbs. The gnu, or wildebeest, eat short grasses.

Page 76—Mountains

Let students know how mountain ranges affect the climate. For example, mountains on the West Coast of the United States block the winds from the Pacific Ocean. These winds carry a great deal of water vapor. So areas to the west of the mountains have a wetter climate than areas to the east, which are protected from the wind.

Page 78—The Arctic

Students may be interested in learning more about lichens. These plants are common on the tundra. Lichens are actually a combination of two kinds of organisms, a fungus and an alga. The fungus stores water and helps to anchor the lichen. The alga provides food through photosynthesis. This association is so successful that lichens can live in the harsh conditions of the tundra. They can even grow on bare rock, producing acids that break down the rock and form soil.

Page 80—Make a Matching Map

This activity helps students develop the process skill of *Making Models* as they model a map to identify various land forms. Display several maps and lead a discussion of the different map keys and their codes.

Mastery Test, Part A, covering units 1-4 can be found on page T15.

UNIT 5 Shaping the Surface

In Unit 5 the effects of water, ice, and wind on Earth are discussed. Features produced by weathering and erosion are described. Finally, the ways that people can change Earth's surface are presented.

Page 82—Weathering and Erosion

Help students distinguish between weathering and erosion. Weathering is the breaking up of rocks and soil. In erosion, these small pieces of material are carried away from their original position. The material is not lost; rather, it is deposited in another place where new land forms are produced.

Page 84—Water Changes the Surface
Materials: nature magazines and books

Have students look through nature magazines to find examples of water erosion around oceans and rivers. They may even be able to find examples of soil erosion of farmland. Then have students research ways water and soil erosion are being controlled along beaches and riverbanks and on farms. Have students share their pictures and research with the class.

Page 86—Ice Changes the Surface

Students may be interested to learn that during Earth's history large areas of land were covered with sheets of ice about 10,000 feet thick. The ice advanced and retreated many times, changing the land as it moved. The last of these ice ages ended about 10,000 years ago.

Page 88—Wind Changes the Surface

One of the most common features created by the wind are dunes. Dunes are hills formed from windblown sand. They are common in desert areas and along beaches. Some dunes may become over 100 feet tall. Dunes are constantly changing shape and can appear to move over the ground as they are reshaped by the wind. Have students who have been to an ocean beach describe any sand dunes they saw. Ask students what they think happens to these dunes as the wind blows. (They may move and change shape.)

Page 90—Soil

Materials: sample of soil, magnifying glass, paper towel

Show students a sample of soil from a garden, park, or area around the school. Spread out the soil on a paper towel. Have students examine the soil with a magnifying glass. Challenge them to find bits of rock, twigs, leaves, and other particles that make up soil.

Page 92—People Change the Surface

Discuss with students how people can help repair the damage to Earth's surface. Examples include planting trees in areas that have been deforested, being careful with campfires, and resurfacing and planting over areas that have been mined.

Page 94—Show Weathering Effects

This activity helps students develop the process skill *Observing* as they demonstrate mountain and plateau erosion. Students can further their observational skills and learn about weathering by examining the school campus for examples of erosion.

UNIT 6 The Changing Earth

Unit 6 looks at how forces change Earth from within. The different layers of Earth are described, and the plates of the crust and their movements are explained. The formation of mountains and volcanoes and the occurrence of earthquakes are described in terms of the motion of the plates. Finally, the concept of geothermal energy is explored, and its potential uses are discussed.

Page 96—The Structure of Earth

Help students understand the diagram of the interior of Earth. (You may first want to review the terms *diameter* and *radius*.) Explain that the radius of Earth, or the distance from the center to the edge, is about 4,000 miles. Then have them find the inner core on the diagram. Point out that the radius of the inner core is 800 miles. The outer core is measured from the edge of the inner core outward. Point out each of the remaining layers. Remind students that the part of Earth we see is the crust. The crust makes up the continents and continues under the ocean and forms the ocean floor.

Page 98—Plates of Earth

Materials: heat-proof glass pot, water, light wooden blocks, food coloring, heat source

Make a model to show students how solid plates of Earth can float and move on hot liquid. Put water in a heat-proof glass pot and float a few wooden blocks in the water. Heat the water to boiling. Add a few drops of food coloring at the edge of the pot. Have students observe the currents in the water. Also have them observe how the blocks move.

Page 100—How Mountains Are Formed

Materials: encyclopedia, world map

Write a list of mountain ranges on the board, such as the Rocky, Appalachian, Sierra Nevada, Himalayan, Tien Shan, Alp, and Andes mountains. Have students divide into groups to research a mountain range. Students should include information about where the mountains are located, the type of mountain most prevalently found, and the highest peak. One representative from each group should report back to the class.

Page 102—Earthquakes

Materials: encyclopedia, newspaper articles

Have students research a recent earthquake. Where was the epicenter of the earthquake? How far from the epicenter was the earthquake felt? How much damage resulted from the earthquake?

Interested students may want to learn about the Richter scale. This scale measures how much energy an earthquake releases. The more energy, the stronger the earthquake.

Page 104—Volcanoes

Materials: small jar, clay, vinegar, baking soda

Students can make their own volcano. Have students build clay up around a jar to make a "mountain." Then have them add a teaspoon of baking soda to the jar. When vinegar is added, the result is an "erupting" volcano.

Page 106—Energy Inside Earth

Old Faithful is one of about 200 geysers in Yellowstone National Park. The eruption of most of these geysers is unpredictable. Ask any students who have visited the park to describe the geysers they saw. If photographs are available, pass them around for students to see.

Page 108—Show How Mountains Form

This activity reinforces the process skill *Classifying* as students experiment with ways different types of mountains are formed. Students can learn about specific mountains by reading library resource materials.

UNIT 7 — Materials of Earth

Unit 7 describes the materials that make up Earth's crust. Minerals and their properties are described. The unit also discusses the kinds of rocks and how they are formed. Materials, like ores, coals, and petroleum, that are useful to people are discussed. The unit also describes fossils and how they are formed.

Page 110—Minerals

Materials: several mineral samples, streak plate (piece of unglazed porcelain)

Tell students that streak is another way in which minerals can be identified. Streak is the color left behind when a mineral is rubbed along a rough surface. A mineral's streak may be different from the mineral's color. Help students find the streak of several samples. Have them make a chart that lists the name of the mineral, its color, and its streak.

Page 112—Igneous Rocks

Materials: samples of obsidian, granite, and basalt; magnifying glass

Have students look at samples of obsidian, granite, and basalt with a magnifying glass. Ask them to determine which ones have crystals and where the crystals are larger. (Granite has larger crystals than basalt. Obsidian has no crystals.)

Page 114—Sedimentary Rocks

Materials: large jar with lid, gravel, coarse sand, soil, water

Have students make a model of the formation of layers of sediment. Fill a jar half full with soil, sand, and gravel. Add water until the jar is about three-quarters full. Put the lid on the jar and shake the jar until the contents are thoroughly mixed. Let the materials settle. Then ask students to identify the material in each layer. (The heaviest materials should settle out first.)

Page 116—Metamorphic Rocks

Help students understand how rocks form and change within the rock cycle. Put this diagram on the chalkboard:

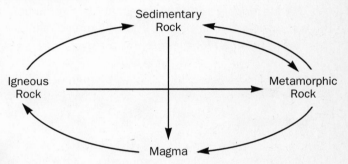

Then label each "arrow," pointing out the process that changes one kind of rock to another. (For example, when weathering breaks down igneous rock, the material may be carried away and deposited as sediments that eventually form sedimentary rock.)

Page 118—Ores

Have students name objects in the home or the classroom that are made from minerals. Objects may contain lead, gold, aluminum, iron, copper, or graphite. Write the name of each object on the chalkboard under the name of the mineral from which it is made.

Page 119—Fossils

Materials: samples of fossils, magnifying glass

Bring in several samples of plant and animal fossils for the students to observe. Encourage students to use a magnifying glass to look at each one. Ask them to describe the plant or animal that formed the fossil. Finally, ask them if they think the organism resembles anything that is alive today.

Page 120—Coal

Materials: encyclopedia

Have students use an encyclopedia to find out where coal deposits are located in the United States. Ask students to determine whether their state has coal and, if so, what type.

Page 122—Petroleum

Emphasize the importance of petroleum in the world today. Products derived from petroleum include fuels for different forms of transportation, oils for heating and lubricating, wax, asphalt for paving roads, drugs, cosmetics, explosives, fertilizers, paint, plastics, and artificial rubber. There are problems involved with the use of petroleum, however. Tell students that they will learn about these problems in Unit 8.

Page 124—Make a Fossil

This activity encourages students to use the process skill *Making Models* as they learn about fossil formation. Students would further benefit if fossils or sedimentary rocks were provided for them to examine.

UNIT 8 Conservation

Unit 8 discusses natural resources and the role of people in conserving these resources. The problems associated with soil erosion and air and water pollution are presented. Ways of lessening these problems are also discussed. Finally, the importance of conservation is stressed.

Page 126—Earth in Balance

Write the words *Renewable Natural Resources* and *Nonrenewable Natural Resources* on the chalkboard. Have students name things people use that would fall into these categories. Write their suggestions under each heading. Have students explain their choices.

Page 128—Saving the Soil

Materials: two tin pans, soil, watering can

Make a hill of soil with sides of about 45 degrees in each tin pan. Make a series of contoured rows on the second hill. Gently pour water from the watering can over the top of each hill. Have students observe what happens to each hill and decide where the run-off carries away more soil.

Page 130—Keeping Air Clean

Materials: small white paper plates, petroleum jelly

Have students make some "Pollution Detectors." Have them coat several paper plates with petroleum jelly. Then place the plates in several locations. You may want to choose a park, a parking lot, and a location near school. Leave the plates for several days to allow time for particles from the air to settle on them. Ask students where they detected the most pollution.

Page 132—Keeping Water Clean

Review the water cycle with students. Help them understand that water can be considered a renewable natural resource because it is not lost from Earth. The water cycle helps to purify water because chemicals in water are left behind when the water evaporates. Water can also be purified in special treatment plants where bacteria and other harmful organisms are killed.

Page 134—Saving Our Resources

Materials: nature magazines, encyclopedia

Have students research to find out what animals are endangered. Then have them find pictures of the animals in magazines. Have them cut out the pictures and make a bulletin board display. Help students determine why the animal is endangered and arrange the display accordingly. For example, you could arrange the pictures under the headings *Loss of Environment, Hunting,* and *Pollution.*

Page 136—Make a Conservation Poster

This activity implements the process skill of *Communicating* as students identify and discuss ways to protect natural resources in their community. Display students' posters in the classroom.

Mastery Test, Part B, covering units 5-8 can be found on page T17.

Mastery Test

Fill in the circle in front of the word or phrase that best completes each sentence. The first one is done for you.

1. The nine objects in orbit around the sun are
   - ⓐ rings.
   - ● planets.
   - ⓒ telescopes.

2. The only planet that has living things as we know them is
   - ⓐ Venus.
   - ⓑ Jupiter.
   - ⓒ Earth.

3. Stars are made of
   - ⓐ cold rock.
   - ⓑ frozen water.
   - ⓒ hot gases.

4. The atmosphere is a mixture of
   - ⓐ gases.
   - ⓑ rocks.
   - ⓒ liquids.

5. The distance north or south of the equator is
   - ⓐ altitude.
   - ⓑ height.
   - ⓒ latitude.

6. Most changes in weather take place along
   - ⓐ oceans.
   - ⓑ fronts.
   - ⓒ poles.

7. The largest land masses on Earth are called
   - ⓐ caves.
   - ⓑ plateaus.
   - ⓒ continents.

8. Three fourths of Earth is covered with
   - ⓐ mountains.
   - ⓑ plains.
   - ⓒ water.

9. Clouds are made of tiny drops of
   - ⓐ nitrogen.
   - ⓑ oxygen.
   - ⓒ water.

10. The average weather of a region over a long period of time is
    - ⓐ tundra.
    - ⓑ climate.
    - ⓒ current.

Fill in the missing words.

11. In a _____ climate, winters are cold and summers are warm. (temperate, tropical)

12. All the planets are held in orbit by a force called _____. (gravity, rotation)

13. More kinds of plants and animals are found in the _____ forest than anywhere else on Earth. (tropical rain, deciduous)

14. Most of the water on Earth is in the _____. (clouds, oceans)

Use the words below to complete the sentences.

atmosphere	precipitation	run-off
orbit	river	

15. The curved path of a planet around the sun is called an _____.

16. Rain, snow, sleet, and hail are all forms of _____.

17. A _____ is a natural flow of water that runs into a lake, ocean, or other body of water.

18. The water cycle is the movement of water between the ground and the

    _____.

19. Rainwater that flows across the ground is called _____.

Write the answer on the lines.

20. Name two factors that affect the climate of an area.

_____

_____

Mastery Test

Fill in the circle in front of the word or phrase that best completes each sentence. The first one is done for you.

1. Gasoline is made from
   - ⓐ coal.
   - ● petroleum.
   - ⓒ magma.

2. Weathering is the
   - ⓐ moving of rocks and soil.
   - ⓑ breaking up of rocks and soil.
   - ⓒ flow of water over rocks.

3. Erosion is the
   - ⓐ moving of rocks and soil.
   - ⓑ breaking up of rocks and soil.
   - ⓒ the flow of water over ground.

4. The major cause of weathering and erosion is
   - ⓐ rich soil.
   - ⓑ sunshine.
   - ⓒ moving water.

5. When the plates of Earth collide, they can form
   - ⓐ volcanoes.
   - ⓑ plains.
   - ⓒ rivers.

6. Most minerals are formed by
   - ⓐ water.
   - ⓑ magma.
   - ⓒ lava.

7. People can help keep air clean by
   - ⓐ riding in car-pools.
   - ⓑ burning fuels.
   - ⓒ putting soot in the air.

8. The crust is made up of about 20
   - ⓐ peaks.
   - ⓑ plates.
   - ⓒ continents.

9. The constant changing of rocks is called
   - ⓐ the rock cycle.
   - ⓑ the heat cycle.
   - ⓒ strata.

10. Burning fossil fuels is the cause of most
    - ⓐ soil erosion.
    - ⓑ endangered species.
    - ⓒ air pollution.

Mastery Test

Write the word or words that best finish each sentence.

11. When water freezes, it takes up _____ space than it does as a liquid.

12. Most earthquakes are caused by the sudden movement of plates along a

    _____.

13. The main cause of water _____ is the release of harmful materials into nearby bodies of water.

14. Layers of sedimentary rock are called _____.

Use the words below to complete the sentences.

| acids | crystals | rocks |
| crust | plates | |

15. The surface of Earth is made up of large sections called _____.

16. When _____ pollute lakes, fish die, and the water is not safe to drink.

17. Soil is made of _____ that slowly break down into tiny bits.

18. The atoms of a mineral are arranged in regular patterns and form shapes

    called _____.

19. When two plates move, the _____ near the fault is squeezed and stretched, causing an earthquake.

Write the answer on the lines.

20. What is the difference between coal and petroleum?

    _____

    _____

Mastery Test

Fill in the circle in front of the word or phrase that best completes each sentence. The first one is done for you.

1. The nine objects in orbit around the sun are
   - (a) rings.
   - ● planets.
   - (c) telescopes.

2. The only planet that has living things as we know them is
   - (a) Venus.
   - (b) Jupiter.
   - (c) Earth. ●

3. Stars are made of
   - (a) cold rock.
   - (b) frozen water.
   - (c) hot gases. ●

4. The atmosphere is a mixture of
   - (a) gases. ●
   - (b) rocks.
   - (c) liquids.

5. The distance north or south of the equator is
   - (a) altitude.
   - (b) height.
   - (c) latitude. ●

6. Most changes in weather take place along
   - (a) oceans.
   - (b) fronts. ●
   - (c) poles.

7. The largest land masses on Earth are called
   - (a) caves.
   - (b) plateaus.
   - (c) continents. ●

8. Three fourths of Earth is covered with
   - (a) mountains.
   - (b) plains.
   - (c) water. ●

9. Clouds are made of tiny drops of
   - (a) nitrogen.
   - (b) oxygen.
   - (c) water. ●

10. The average weather of a region over a long period of time is
    - (a) tundra.
    - (b) climate. ●
    - (c) current.

Mastery Test

Fill in the missing words.

11. In a _____temperate_____ climate, winters are cold and summers are warm. (temperate, tropical)

12. All the planets are held in orbit by a force called _____gravity_____. (gravity, rotation)

13. More kinds of plants and animals are found in the _____tropical rain_____ forest than anywhere else on Earth. (tropical rain, deciduous)

14. Most of the water on Earth is in the _____oceans_____. (clouds, oceans)

Use the words below to complete the sentences.

atmosphere	precipitation	run-off
orbit	river	

15. The curved path of a planet around the sun is called an _____orbit_____.

16. Rain, snow, sleet, and hail are all forms of _____precipitation_____.

17. A _____river_____ is a natural flow of water that runs into a lake, ocean, or other body of water.

18. The water cycle is the movement of water between the ground and the _____atmosphere_____.

19. Rainwater that flows across the ground is called _____run-off_____.

Write the answer on the lines.

20. Name two factors that affect the climate of an area.
    _____Two factors that affect the climate of an area_____
    _____are latitude and nearness to ocean currents._____

Mastery Test

Fill in the circle in front of the word or phrase that best completes each sentence. The first one is done for you.

1. Gasoline is made from
   - (a) coal.
   - ● petroleum.
   - (c) magma.

2. Weathering is the
   - (a) moving of rocks and soil.
   - (b) breaking up of rocks and soil. ●
   - (c) flow of water over rocks.

3. Erosion is the
   - (a) moving of rocks and soil. ●
   - (b) breaking up of rocks and soil.
   - (c) the flow of water over ground.

4. The major cause of weathering and erosion is
   - (a) rich soil.
   - (b) sunshine.
   - (c) moving water. ●

5. When the plates of Earth collide, they can form
   - (a) volcanoes. ●
   - (b) plains.
   - (c) rivers.

6. Most minerals are formed by
   - (a) water.
   - (b) magma. ●
   - (c) lava.

7. People can help keep air clean by
   - (a) riding in car-pools. ●
   - (b) burning fuels.
   - (c) putting soot in the air.

8. The crust is made up of about 20
   - (a) peaks.
   - (b) plates. ●
   - (c) continents.

9. The constant changing of rocks is called
   - (a) the rock cycle. ●
   - (b) the heat cycle.
   - (c) strata.

10. Burning fossil fuels is the cause of most
    - (a) soil erosion.
    - (b) endangered species.
    - (c) air pollution. ●

Mastery Test

Write the word or words that best finish each sentence.

11. When water freezes, it takes up _____more_____ space than it does as a liquid.

12. Most earthquakes are caused by the sudden movement of plates along a _____fault_____.

13. The main cause of water _____pollution_____ is the release of harmful materials into nearby bodies of water.

14. Layers of sedimentary rock are called _____strata_____.

Use the words below to complete the sentences.

acids	crystals	rocks
crust	plates	

15. The surface of Earth is made up of large sections called _____plates_____.

16. When _____acids_____ pollute lakes, fish die, and the water is not safe to drink.

17. Soil is made of _____rocks_____ that slowly break down into tiny bits.

18. The atoms of a mineral are arranged in regular patterns and form shapes called _____crystals_____.

19. When two plates move, the _____crust_____ near the fault is squeezed and stretched, causing an earthquake.

Write the answer on the lines.

20. What is the difference between coal and petroleum?
    _____Coal is a solid fossil fuel,_____
    _____and petroleum is a liquid fossil fuel._____

Just for Fun

Each riddle below is about the solar system and beyond. Solve each riddle. Choose from the words below. Write the words on the spaces.

asteroid	moon	stars
Jupiter	Polaris	Venus
Mars	Saturn	Voyager

1. I am made of rocks and metal. I travel around the sun.

   I am an _____.

2. We are made of hot gases. We give off heat and light.

   We are _____.

3. I travel in an orbit around Earth. I reflect light from the sun.

   I am Earth's _____.

4. I am a star. When you are looking at me, you are facing north.

   My name is _____.

5. I am the second planet from the sun. I am about the same size as Earth.

   My name is _____.

6. I am the largest planet in the solar system. My orbit takes 12 Earth years.

   My name is _____.

7. I am covered with red dust. I am often called the Red Planet.

   My name is _____.

8. I am known for the shiny rings around me.

   My name is _____.

9. I am a space probe. I took pictures of the outer planets.

   My name is _____.

Just for Fun

Use the clues to complete the puzzle. Choose from the words below.

atmosphere	gravity	thermometer
cirrus	lightning	tornado
condensation	nitrogen	water vapor

Across

2. Clouds that are white and feathery
4. An instrument used to measure temperature
5. When water vapor in the air changes back to a liquid
7. The force that holds the gases of the atmosphere around Earth
8. A violent but small storm
9. Water as a gas

Down

1. Caused by electric charges
3. The blanket of air around Earth
6. Three fourths of the atmosphere

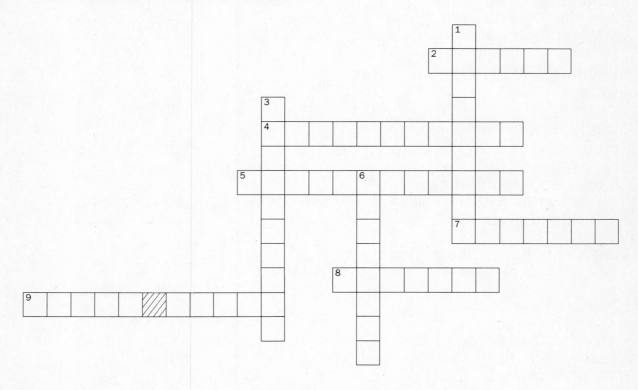

Just for Fun

All the statements below have something to do with land and water. Match each statement with one of the words below. Use the letters in LAND AND WATER to help you.

abyssal plain	limestone	trenches
continents	plain	tributary
ground water	ridge	valleys
lake	stalactite	wells

1. lies in a basin   **L** _ _ _

2. part of the ocean floor   _ _ _ _ _ **A** _

3. largest land masses on Earth   _ _ **N** _ _ _ _ _ _

4. where two sloping surfaces meet   _ _ **D** _ _

5. river that flows into another river   _ _ _ _ _ _ _ **A** _

6. large area of flat land   _ _ _ _ **N**

7. underground water   _ _ _ _ _ _ **D** _ _ _ _ _

8. holes dug to get ground water   **W** _ _ _

9. low places between mountains   _ **A** _ _ _ _ _

10. hangs from roofs of caves   _ **T** _ _ _ _ _ _ _

11. rock found in most caves   _ _ _ **E** _ _ _ _ _

12. deep cuts in the ocean floor   **R** _ _ _ _ _ _

Just for Fun

Use the clues to complete the puzzle. Choose from the words below.

Arctic	grasslands	redwoods
cactus	latitude	temperate
climate	mountain	tropical
deserts	polar	tundra

Across

1. places where many kinds of grasses grow
4. distance north or south of the equator
5. the climate zone of the United States
8. a climate that is warm all year round
9. places where less than 10 inches of rain fall in a year
11. a landform that may have several different climate zones from bottom to top
12. a kind of plant in a desert

Down

2. the land around the North Pole
3. the average weather of a region over a long period of time
6. a climate that is cold most of the year
7. trees that grow in evergreen forests
10. an area in the Arctic where no trees grow

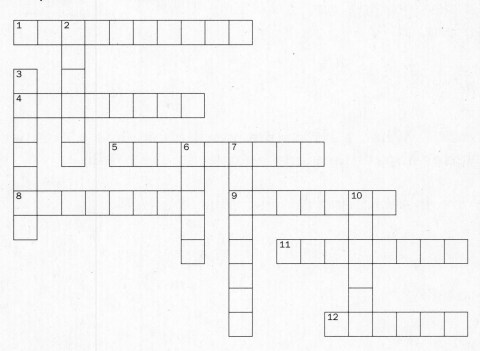

Just for Fun

Match each description about weathering and erosion with one of the words below. Then write the words on the spaces at the right.

glacier	rainfall	valley
mining	river	weathering
ocean	snow	wind

1. the breaking up of rocks and soil   _ _ _ _ _ _ _ _ _ _ _
                                                                    3

2. carries away bits of rock and soil as it flows downhill   _ _ _ _ _
                                                                           4

3. soft flakes of frozen water that fall to Earth   _ _ _ _
                                                                8

4. can damage buildings and push over large trees   _ _ _ _
                                                                   2

5. large, slow-moving sheet of ice   _ _ _ _ _ _ _
                                                         7

6. has waves that pound against the shore   _ _ _ _ _
                                                            6

7. some soaks into the soil and some flows over the ground   _ _ _ _ _ _ _ _
                                                                   5

8. way of getting coal   _ _ _ _ _ _
                                 1

Solve the riddle. When a letter of a word above has a number under it, write that letter above the same number in the riddle.

Riddle: We are in rocks and humus. What are we?

_ _ _ _ _ _ _ _
1 2 3 4 5 6 7 8

Just for Fun

Each sentence below is about the changing Earth. But a word is missing from each sentence. Use the words below to complete the sentences. Then write the words in the spaces at the right. The first one is done for you.

cinders	erupt	iron
dome mountains	fold mountains	mantle
earthquake	hot springs	plates

1. Earth's inner core is made of nickel and   <u>I</u> <u>R</u> <u>O</u> <u>N</u> .

                                                            4

2. Steam and hot water escape in   _ _ _   _ _ _ _ _ _ _ _ _ .

     7

3. Shaking of Earth's crust is an   _ _ _ _ _ _ _ _ _ _ .

     8

4. The thickest layer of Earth is the   _ _ _ _ _ _ .

     6

5. Earth's crust is made of 20   _ _ _ _ _ _ .

     2

6. A bulge in a plate forms   _ _ _ _   _ _ _ _ _ _ _ _ _ .

     5

7. Volcanic rocks the size of golf balls are   _ _ _ _ _ _ _ .

8. Lava, gas, and rocks pour out when volcanoes   _ _ _ _ _ .

     3

9. Folds in Earth's plates form   _ _ _ _   _ _ _ _ _ _ _ _ _ .

     1

Solve the riddle. When a letter of a word above has a number under it, write that letter above the same number in the riddle.

Riddle: I am a famous volcano. You can find me in Hawaii. But be careful. I'm still an active volcano. Who am I?

_ _ _ _ _   _ _ _
1 2 3 4 5   6 7 8

Just for Fun

Use the clues to complete the puzzle. Choose from the words below.

aluminum	granite	oil
basalt	metals	pumice
chalk	minerals	rocks

Across

3. These solid substances form in nature. Rocks are made of them.

6. This is a sedimentary rock made from seashells and minerals.

7. This is a liquid fossil fuel.

8. This is an igneous rock that floats.

Down

1. This is a common igneous rock often used for building.

2. This igneous rock forms under the ocean and makes pillow-shaped lumps.

3. These are minerals that can transfer heat and electricity.

4. This metal is used to make airplanes.

5. Igneous, sedimentary, and metamorphic are the three kinds of _____.

© 1996 Steck-Vaughn Company, *The Earth and Beyond*, Just for Fun, Unit 7.

Just for Fun

Match each description about conservation with one of the words below. Then write the words on the spaces at the right.

acid rain	conservation	refuges
air pollution	fuels	sewage
chemicals	reclaimed	smog

1. soil may be put back on Earth's surface, or

    _ _ _ _ _ _ _ _ _
    1

2. places where plants and animals are protected

    _ _ _ _ _ _ _
    2

3. air pollution that hangs over cities

    _ _ _ _
    3

4. dirty air

    _ _ _   _ _ _ _ _ _ _ _ _
       4

5. causes most air pollution

    _ _ _ _ _
       5

6. mix of air pollution and moisture

    _ _ _ _   _ _ _ _
          6

7. dumped in waterways by factories

    _ _ _ _ _ _ _ _
    7

8. human wastes

    _ _ _ _ _ _
       8

9. wise use of Earth's resources

    _ _ _ _ _ _ _ _ _ _ _ _
          9

Solve the riddle. When the letter of a word above has a number under it, write that letter above the same number in the riddle.

Riddle: We are in Earth naturally. What are we?

_ _ _ _ _ _ _ _ _
1 2 3 4 5 6 7 8 9

UNIT 1   Just for Fun

Each riddle below is about the solar system and beyond. Solve each riddle. Choose from the words below. Write the words on the spaces.

asteroid	moon	stars
Jupiter	Polaris	Venus
Mars	Saturn	Voyager

1. I am made of rocks and metal. I travel around the sun.

   I am an A S T E R O I D .

2. We are made of hot gases. We give off heat and light.

   We are S T A R S .

3. I travel in an orbit around Earth. I reflect light from the sun.

   I am Earth's M O O N .

4. I am a star. When you are looking at me, you are facing north.

   My name is P O L A R I S .

5. I am the second planet from the sun. I am about the same size as Earth.

   My name is V E N U S .

6. I am the largest planet in the solar system. My orbit takes 12 Earth years.

   My name is J U P I T E R .

7. I am covered with red dust. I am often called the Red Planet.

   My name is M A R S .

8. I am known for the shiny rings around me.

   My name is S A T U R N .

9. I am a space probe. I took pictures of the outer planets.

   My name is V O Y A G E R .

UNIT 2   Just for Fun

Use the clues to complete the puzzle. Choose from the words below.

atmosphere	gravity	thermometer
cirrus	lightning	tornado
condensation	nitrogen	water vapor

Across
2. Clouds that are white and feathery
4. An instrument used to measure temperature
5. When water vapor in the air changes back to a liquid
7. The force that holds the gases of the atmosphere around Earth
8. A violent but small storm
9. Water as a gas

Down
1. Caused by electric charges
3. The blanket of air around Earth
6. Three fourths of the atmosphere

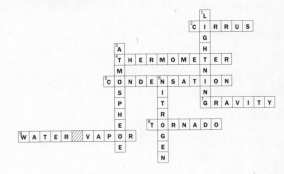

UNIT 3   Just for Fun

All the statements below have something to do with land and water. Match each statement with one of the words below. Use the letters in LAND AND WATER to help you.

abyssal plain	limestone	trenches
continents	plain	tributary
ground water	ridge	valleys
lake	stalactite	wells

1. lies in a basin

2. part of the ocean floor

3. largest land masses on Earth

4. where two sloping surfaces meet

5. river that flows into another river

6. large area of flat land

7. underground water

8. holes dug to get ground water

9. low places between mountains

10. hangs from roofs of caves

11. rock found in most caves

12. deep cuts in the ocean floor

L A K E
A B Y S S A L   P L A I N
C O N T I N E N T S
R I D G E
T R I B U T A R Y
P L A I N
G R O U N D   W A T E R
W E L L S
V A L L E Y S
S T A L A C T I T E
L I M E S T O N E
T R E N C H E S

UNIT 4   Just for Fun

Use the clues to complete the puzzle. Choose from the words below.

Arctic	grasslands	redwoods
cactus	latitude	temperate
climate	mountain	tropical
deserts	polar	tundra

Across
1. places where many kinds of grasses grow
4. distance north or south of the equator
5. the climate zone of the United States
8. a climate that is warm all year round
9. places where less than 10 inches of rain fall in a year
11. a landform that may have several different climate zones from bottom to top
12. a kind of plant in a desert

Down
2. the land around the North Pole
3. the average weather of a region over a long period of time
6. a climate that is cold most of the year
7. trees that grow in evergreen forests
10. an area in the Arctic where no trees grow

UNIT 5 Just for Fun

Match each description about weathering and erosion with one of the words below. Then write the words on the spaces at the right.

glacier	rainfall	valley
mining	river	weathering
ocean	snow	wind

1. the breaking up of rocks and soil W E A T H E R I N G
                                                      3

2. carries away bits of rock and soil as it flows downhill R I V E R
                                                                    4

3. soft flakes of frozen water that fall to Earth S N O W
                                                          8

4. can damage buildings and push over large trees W I N D
                                                          2

5. large, slow-moving sheet of ice G L A C I E R
                                            7

6. has waves that pound against the shore O C E A N
                                                  6

7. some soaks into the soil and some flows over the ground R A I N F A L L
                                                                  5

8. way of getting coal M I N I N G
                              1

Solve the riddle. When a letter of a word above has a number under it, write that letter above the same number in the riddle.

Riddle: We are in rocks and humus. What are we?

M I N E R A L S
1 2 3 4 5 6 7 8

UNIT 6 Just for Fun

Each sentence below is about the changing Earth. But a word is missing from each sentence. Use the words below to complete the sentences. Then write the words in the spaces at the right. The first one is done for you.

cinders	erupt	iron
dome mountains	fold mountains	mantle
earthquake	hot springs	plates

1. Earth's inner core is made of nickel and I R O N .
                                                    4

2. Steam and hot water escape in H O T   S P R I N G S
                                        7

3. Shaking of Earth's crust is an E A R T H Q U A K E .
                                                  8

4. The thickest layer of Earth is the M A N T L E .
                                              6

5. Earth's crust is made of 20 P L A T E S .
                                      2

6. A bulge in a plate forms D O M E   M O U N T A I N S .
                                                5

7. Volcanic rocks the size of golf balls are C I N D E R S .

8. Lava, gas, and rocks pour out when volcanoes E R U P T .
                                                        3

9. Folds in Earth's plates form F O L D   M O U N T A I N S .
                                                    1

Solve the riddle. When a letter of a word above has a number under it, write that letter above the same number in the riddle.

Riddle: I am a famous volcano. You can find me in Hawaii. But be careful. I'm still an active volcano. Who am I?

M A U N A   L O A
1 2 3 4 5   6 7 8

UNIT 7 Just for Fun

Use the clues to complete the puzzle. Choose from the words below.

aluminum	granite	oil
basalt	metals	pumice
chalk	minerals	rocks

Across

3. These solid substances form in nature. Rocks are made of them.

6. This is a sedimentary rock made from seashells and minerals.

7. This is a liquid fossil fuel.

8. This is an igneous rock that floats.

Down

1. This is a common igneous rock often used for building.

2. This igneous rock forms under the ocean and makes pillow-shaped lumps.

3. These are minerals that can transfer heat and electricity.

4. This metal is used to make airplanes.

5. Igneous, sedimentary, and metamorphic are the three kinds of _____.

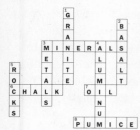 

UNIT 8 Just for Fun

Match each description about conservation with one of the words below. Then write the words on the spaces at the right.

acid rain	conservation	refuges
air pollution	fuels	sewage
chemicals	reclaimed	smog

1. soil may be put back on Earth's surface, or R E C L A I M E D
                                                        1

2. places where plants and animals are protected R E F U G E S
                                                          2

3. air pollution that hangs over cities S M O G
                                                3

4. dirty air A I R   P O L L U T I O N
                                  4

5. causes most air pollution F U E L S
                                      5

6. mix of air pollution and moisture A C I D   R A I N
                                                      6

7. dumped in waterways by factories C H E M I C A L S
                                              7

8. human wastes S E W A G E
                          8

9. wise use of Earth's resources C O N S E R V A T I O N
                                                          9

Solve the riddle. When the letter of a word above has a number under it, write that letter above the same number in the riddle.

Riddle: We are in Earth naturally. What are we?

R E S O U R C E S
1 2 3 4 5 6 7 8 9

Special Projects

These Special Projects are motivational activities that supplement and provide enrichment to each unit. Read through the projects before beginning a unit to determine when each is most helpful to your class.

UNIT 1 — The Solar System

Earth's Moon
Materials: encyclopedia or book on space exploration

Interested students may want to research the moon. Some students can learn about conditions on the moon: if there is water, if there is an atmosphere, what kinds of features the moon has. Other students may want to write a report describing the Apollo missions to the moon. Have volunteers from each group present their reports to the class.

Sunlight on Earth
Materials: graph paper, flashlight

Help students visualize how sunlight differs at different places on Earth. Shine a flashlight directly on a sheet of graph paper. Have a student draw a circle around the lighted area. Tell students that the circle stands for the area of Earth that receives the direct rays of the sun. Now, using a second sheet of graph paper, shine the flashlight on the paper at an angle. Have a student trace the area on the paper. Help students see that the light rays that strike at an angle spread out over a greater area. Areas where the rays are more spread out are colder than areas where rays strike directly.

UNIT 2 — Weather

The Greenhouse Effect
Materials: two thermometers, glass bowl

Tell students that carbon dioxide in the atmosphere helps to trap heat around Earth. This process is called the greenhouse effect. Make a model to illustrate this effect. Place two thermometers in the sun. Place a glass bowl over one thermometer. Explain that the glass bowl acts much the same as the atmosphere does to trap heat. Have students take temperature readings from both thermometers every 10 minutes for an hour. Then ask students which has the higher temperature.

Lightning Safety
Materials: encyclopedia, poster paper, marking pens

Have students research how people can protect themselves from lightning during a thunderstorm.

Students may also want to illustrate these guidelines for a bulletin board display.

UNIT 3 — Land and Water

Local Features
Materials: local map (such as a road map)

Have students look at a map of the area in which they live. Have them identify major land features in the area. Make a chart with the headings: *Lakes, Ponds, Rivers,* and *Mountains.* Give the names of local examples. If your area is near the ocean or near major caves, include these headings in your chart.

Where Does the Water Come From?

If you live in an area that gets its water from wells, you may want to ask a local well digger to speak to the class. Ask the students to be prepared with a short list of questions. They may want to ask how deep the wells are and whether they ever run dry.

In other areas, contact the local water company and ask about the source of water for your city or town. The water company may also be able to provide information on water use–for example, how much water a person uses in a day.

UNIT 4 — Regions of Earth

The Grasslands in History
Materials: book on American history

Have students research to find out what happened to the bison and other animals of the grasslands when settlers moved into the western United States. A student volunteer can make a report on the findings to the class.

Environment Bulletin Board
Materials: magazines, encyclopedia, field guides

Divide the class into five groups. Have each group research one of these environments: the desert, the forests, the grasslands, the mountains, and the tundra. Have students collect pictures of these environments for a bulletin board display. Also have them trace or cut out pictures of some of the plants and animals of the environments and arrange them on the bulletin board according to environment.

UNIT 5   Shaping the Surface

How Earth Changes

Materials: nature magazines, poster paper, marking pens

Have students make a bulletin board display showing the effects of weathering and erosion on Earth. They can illustrate the display with pictures cut from nature magazines. Suggest that they look for pictures of canyons, glaciers, sand dunes, and ocean cliffs. They may also be able to find less dramatic examples, perhaps pictures of gullies in farmland or trees bent by wind.

The Ice Ages

Materials: encyclopedia, atlas, colored pencils, tracing paper

Have students research the last great ice age, which ended about 10,000 years ago. Students can trace a map of North America and color the largest area covered by glaciers during this period. Have students determine whether their state was covered by ice during the ice age.

UNIT 6   The Changing Earth

Mount St. Helens

Materials: encyclopedia, current books on volcanoes

Have students research the eruption of Mount St. Helens in the state of Washington in 1980. Ask students to find out what kinds of materials erupted from the volcano. Ask them to also find out what effects the eruption had on the nearby area. Have a student make a report to the class. The student should be prepared to answer a few questions. Are there other volcanoes in Washington State? Will Mount St. Helens erupt again?

Energy Inside Earth

Materials: encyclopedia

Have students find out how geothermal energy is being used. (It can be used directly to heat homes and water. It can also be used to produce electricity.) Also ask them if any place besides Iceland uses geothermal energy. (Geothermal energy is in use in an area of Italy, in northern California, and in an area of New Zealand.)

UNIT 7   Materials of Earth

Growing Crystals

Materials: glass jar, salt, string, spoon, water

Have students grow salt crystals. Pour a small amount of salt into a jar of hot water. Stir the mixture and continue adding salt until no more will dissolve. Place a string in the solution. Leave the jar in a place where it won't be disturbed. After a few days, have students observe the salt crystals that have formed on the string.

Identifying Rocks

Materials: field guide for rocks or encyclopedia, magnifying glass

Have students bring in samples of rocks. Have them use a field guide or encyclopedia to try to identify their samples. Have students group and label their samples for a classroom display.

UNIT 8   Conservation

Environmental Scrapbook

Materials: recent newspaper and magazine articles on pollution, notebook

Collect any current articles on pollution from newspapers and magazines. Summarize each article for the students. What type of pollution does the article describe? Is this a new problem or one that has been developing for a long time? Does the article offer any solution to the problem? You may want to save the articles in a notebook.

Natural Resources Information

Ask a member of a local environmental organization, federal or state environmental official, or park ranger to speak to the students on the importance of conservation. Find out if any conservation projects are taking place in your area.

Related Steck-Vaughn Products

Additional resources are available from Steck-Vaughn to enrich the content of **The Earth and Beyond.** Many of the titles are available in hard cover or soft cover. Readability range is from 4th to 7th grade level.

New View Series
Planet Earth
Seas and Oceans
Stars and Planets

Raintree Illustrated Science Encyclopedia

World Disasters Series
Earthquake
Flood
Storm
Volcano

Facing the Future Series
Space Science

How Do We Know? Series
About the Universe
The Earth Is Round

Atlas of the Environment

Beginnings—Origins and Evolution Series
The Earth
The Universe

Science Spotlight Series
Astronomy

Wonders of the World Series
Andes Mountains
Antarctica
Grand Canyon

1,000 Facts About Series
The Earth
Space

STECK-VAUGHN ®
C O M P A N Y
ELEMENTARY • SECONDARY • ADULT • LIBRARY

For ordering information call
Steck-Vaughn customer service
1-800-531-5015.

STECK-VAUGHN

WONDERS OF SCIENCE

Joan S. Gottlieb

The Earth and Beyond

ISBN 0-8114-7490-9

5 6 7 8 9 10 PO 99 98

STECK-VAUGHN®
COMPANY
ELEMENTARY • SECONDARY • ADULT • LIBRARY

Contents

UNIT 1 The Solar System

The Parts of the Solar System 4
The Sun and Other Stars 6
Patterns of Stars............................8
The Inner Planets 10
Earth...12
Earth's Moon 14
The Outer Planets 16
Exploring Space 18
A Year 20
Day and Night 21
The Seasons.................................22
Review 23
Explore & Discover
    Model the Seasons 24
Unit Test 25

UNIT 2 Weather

Weather and the Atmosphere...........26
The Water Cycle 28
Clouds...30
Precipitation.................................32
Humidity 34
Temperature 35
What Makes the Weather Change?...36
Air Pressure.................................38
Storms 40
Tracking the Weather.....................42
Review 43
Explore & Discover
    Record Temperatures..................44
Unit Test 45

UNIT 3 Land and Water

Rivers...46
Lakes and Ponds 48
Wells 50
Caves..52
The Oceans 54
Along the Ocean Floor 56
Life in the Oceans.........................58
Continents...................................60
Mountains 62
Plains and Plateaus.......................64
Review 65
Explore & Discover
    Make a Model Island 66
Unit Test 67

UNIT 4 Regions of Earth

Climate Zones...............................68
Deserts 70
Forests 72
Grasslands 74
Mountains 76
The Arctic 78
Review 79
Explore & Discover
    Make a Matching Map 80
Unit Test 81

UNIT 5　Shaping the Surface

Weathering and Erosion 82
Water Changes the Surface 84
Ice Changes the Surface 86
Wind Changes the Surface 88
Soil 90
People Change the Surface 92
Review 93
Explore & Discover
　　Show Weathering Effects 94
Unit Test 95

UNIT 6　The Changing Earth

The Structure of Earth 96
Plates of Earth...............................98
How Mountains Are Formed 100
Earthquakes 102
Volcanoes 104
Energy Inside Earth 106
Review 107
Explore & Discover
　　Show How Mountains Form 108
Unit Test 109

UNIT 7　Materials of Earth

Minerals 110
Igneous Rocks 112
Sedimentary Rocks 114
Metamorphic Rocks......................116
Ores...118
Fossils 119
Coal 120
Petroleum 122
Review 123
Explore & Discover
　　Make a Fossil...........................124
Unit Test 125

UNIT 8　Conservation

Earth in Balance 126
Saving the Soil.............................128
Keeping Air Clean.........................130
Keeping Water Clean 132
Saving Our Resources 134
Review 135
Explore & Discover
　　Make a Conservation Poster 136
Unit Test 137

GLOSSARY 138

The Parts of the Solar System

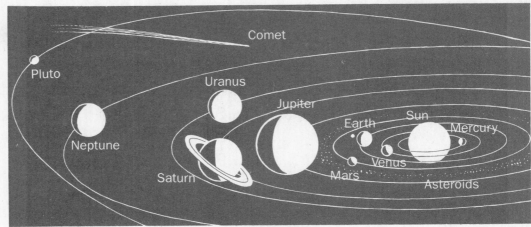

The Solar System

We live on the **planet** Earth. A planet is a large, solid object that travels around a star. Earth is one of nine planets circling a star we call the sun. The sun and the nine planets make up our **solar system**.

Each of the nine planets moves in an **orbit,** or curved path, around the sun. The size of a planet's orbit depends on the distance of the planet from the sun. Planets far from the sun have large orbits. Closer planets have smaller orbits. All the planets are held in orbit by a force called **gravity.** Gravity works like a pull between the sun and the planets. Without gravity, the planets would drift into space.

Some planets have smaller objects called **moons** in orbit around them. Each moon is held in orbit by the gravity between the moon and its planet. There are more than 50 moons orbiting the planets of the solar system. One moon orbits Earth.

Asteroids and **comets** are smaller objects in the solar system. Asteroids are like tiny planets made of rock and metal. They orbit the sun in a belt between the planets Mars and Jupiter.

Comets are made of ice, gas, and dust. A comet has a bright fiery head and a long tail of light. Comets have very large orbits. They travel to the edge of the solar system.

A. Answer True or False.

1. A planet is a large, solid object that travels around a star.

   _____True_____

2. All the planets are held in orbit by a force called gravity. _____True_____

3. Earth has 40 moons. _____False_____

4. Asteroids are made of gas. _____False_____

5. Comets are found between the planets Mars and Jupiter. _____False_____

B. Fill in the missing words.

1. The sun and the nine planets make up the _____solar system_____.
   (solar system, asteroids)

2. Each of the nine planets moves in a curved path around the sun

   called _____an orbit_____. (a comet, an orbit)

3. Planets far from the sun have _____large_____ orbits. (small, large)

4. Each moon is held in orbit around a planet by the gravity between

   the moon and its _____planet_____. (planet, asteroid)

5. There are more than _____50_____ moons in the solar system. (9, 50)

C. Write the names of the planets. Use the diagram on page 4.

_____Mercury, Venus, Earth, Mars, Jupiter,_____

_____Saturn, Uranus, Neptune, Pluto_____

D. Answer the questions.

1. What does the size of the orbit of a planet depend on? _____The size of_____
   _____a planet's orbit depends on the distance of the planet from the sun._____

2. How are all the planets held in orbit? _____All the planets are held_____
   _____in orbit by a force called gravity._____

5

The Sun and Other Stars

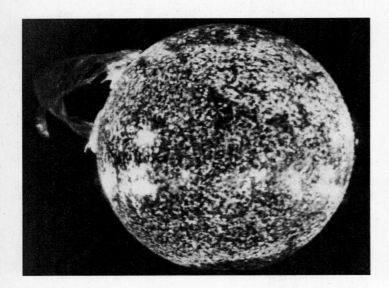

The Sun

If you look at the sky at night, you can see thousands of **stars.** Actually, there are billions of stars, but most of them are too far away to see.

Stars are huge balls of hot gases. They are made up mostly of hydrogen and helium. The center of a star is so hot that the star gives off light. This light energy makes a star shine.

The sun is the star nearest Earth. Yet the sun is 93 million miles away. Because of this distance, it takes 8 minutes for the sun's light to reach Earth.

Other stars are so far away that they look like tiny points of light. Although they look small, most stars are much larger than Earth. The sun is only a medium-size star. But the sun's **diameter,** or distance from one side to the other through the center, is 100 times the diameter of Earth.

The sun is very important to Earth. Plants and animals need the sun's heat and light to stay alive. There would be no life on Earth without the sun.

The sun and other stars are different from planets. Stars give off heat and light. Planets do not. Planets and moons **reflect,** or send back, the light from the sun. Just as a mirror reflects light from a flashlight, planets and moons reflect light from the sun.

A. Fill in the missing words.

1. Stars are huge balls of hot ___gases___ . (gases, rock)

2. The center of a star is so hot that the star gives off ___light___ . (ice, light)

3. The ___sun___ is the star that is nearest Earth. (Jupiter, sun)

4. The sun is a ___medium-size___ star. (medium-size, large)

5. The sun's diameter is ___100___ times the diameter of Earth. (10, 100)

B. Answer True or False.

1. Some stars are too far away for us to see. ___True___

2. Stars are made up mostly of hydrogen and helium. ___True___

3. Most stars are nearer to Earth than the sun. ___False___

4. It takes 8 hours for the sun's light to reach Earth. ___False___

5. Planets give off heat and light. ___False___

C. Answer the questions.

1. What are stars? ___Stars are huge balls of hot gases.___

2. Why does a star give off light? ___The center of a star is so hot that the star gives off light.___

3. How many stars are there? ___There are billions of stars.___

4. Why is the sun important to Earth? ___There would be no life on Earth without the sun.___

5. How are stars different from planets? ___Stars give off heat and light. Planets do not. Planets and moons reflect light from the sun.___

Patterns of Stars

Ursa Minor

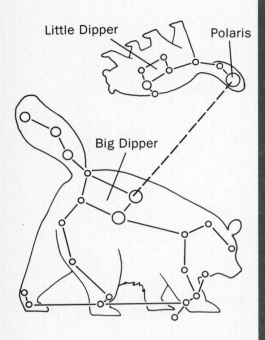

Little Dipper

Polaris

Big Dipper

Ursa Major

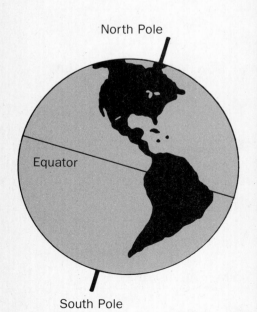

North Pole

Equator

South Pole

People have always looked at and wondered about the stars. Ancient people saw patterns in certain groups of stars. The patterns looked like people, animals, and objects. These groups of stars are called **constellations.** Throughout history, people have used the position of these stars to guide them as they traveled on land and sea.

You may already know some of the 88 constellations. Have you ever seen the Big Dipper? Two stars in the bowl of the Big Dipper point to Polaris, the North Star. Wherever you are, when you look at Polaris, you are facing north. The Big Dipper is part of a large constellation called Ursa Major. Ursa Major is a Latin name that means Great Bear. Polaris is part of Ursa Minor, or the Little Bear.

Every constellation has a Latin name. Many constellations are named after animals. Leo is a lion, Taurus is a bull, and Pisces represents two fish.

At the **North Pole,** the northernmost point on Earth, you can see only the constellations in the northern part of the sky. At the **South Pole,** or southernmost point on Earth, you can see only the constellations in the southern part of the sky. At the **equator,** the imaginary line around the center of Earth, you can see all the constellations.

Because Earth travels around the sun, the constellations appear in different parts of the sky at different times of the year. Some constellations can be seen only during certain seasons.

A. **Circle the letter of the correct answer.**

1. What is a pattern of stars called?

   (a) a constellation   (b) a planet   (c) a moon

2. What do two of the stars in the bowl of the Big Dipper point to?

   (a) Leo   (b) Polaris   (c) Earth

3. When you look at Polaris, in what direction are you facing?

   (a) north   (b) south   (c) west

4. How many constellations are there?

   (a) billions   (b) 24   (c) 88

B. **Answer True or False.**

1. Many constellations are named after animals. __True__

2. Ursa Major is a Latin name that means Great Bear. __True__

3. Each constellation has a French name. __False__

4. At the equator, you can see all of the constellations. __True__

5. Some constellations can be seen only during certain seasons.

   __True__

C. **Answer the questions.**

1. What constellation is the Big Dipper part of? __The Big Dipper is part of a large constellation called Ursa Major.__

2. How have people used stars throughout history? __Throughout history, people have used the positions of the stars to guide them as they traveled on land and sea.__

3. Which constellations can you see if you are at the South Pole? __At the South Pole, you can see only the constellations in the southern part of the sky.__

The Inner Planets

Mercury

Venus

Earth

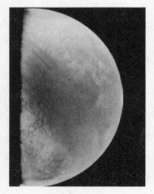

Mars

The four planets that are closest to the sun are called the **inner planets.** The inner planets are Mercury, Venus, Earth, and Mars.

Mercury is the closest planet to the sun. It takes Mercury only 88 days to orbit once around the sun. The surface of Mercury has plains, cliffs, and large holes called <u>craters</u>. Mercury is also one of the smallest planets in the solar system.

Venus is the second planet from the sun. It takes 225 days for Venus to travel once around the sun. Venus is about the same size as Earth. The thick clouds around Venus trap the sun's heat like a greenhouse. So the temperature on Venus can be seven times hotter than the hottest day on Earth!

Earth is the third planet from the sun. It takes Earth about 365 days to orbit the sun once. Earth has one moon. It has oceans of water. It is the only planet with living things as we know them. A special mixture of gases around Earth makes up the **atmosphere.** Plants and animals need the water and the atmosphere to stay alive.

Mars is the fourth planet from the sun. It takes Mars 687 days to make one orbit around the sun. Mars has two moons. Mars is often called the Red Planet because it is covered with a layer of red dust. Scientists think that much of Mars was once covered by water. Today, there is no liquid water on the planet. But the poles are covered with caps that may be made of frozen water.

A. Fill in the missing words.

1. The four planets that are closest to the sun are called the
   ___inner planets___. (inner planets, outer planets)

2. Earth is the ___third___ planet from the sun. (first, third)

3. It takes ___225___ days for Venus to travel once around the sun.
   (88, 225)

4. Earth is the only planet with ___living things___.
   (clouds, living things)

5. The surface of Mercury has plains, cliffs, and ___craters___.
   (rivers, craters)

6. Plants and animals on Earth need the water and
   ___atmosphere___ to stay alive. (atmosphere, red dust)

B. Answer True or False.

1. Venus is called the Red Planet. ___False___

2. Mars is the closest planet to the sun. ___False___

3. Mercury is one of the smallest planets in the solar system.
   ___True___

4. Mars is covered with a layer of red dust. ___True___

5. Thick clouds cover Venus. ___True___

6. Mars has two moons in orbit around it. ___True___

C. Answer the questions.

1. What are the names of the inner planets? ___The inner planets are___
   ___Mercury, Venus, Earth, and Mars.___

2. Earth has water. What other planet do scientists think had water at
   one time? ___Scientists think that much of Mars was once___
   ___covered by water.___

Earth

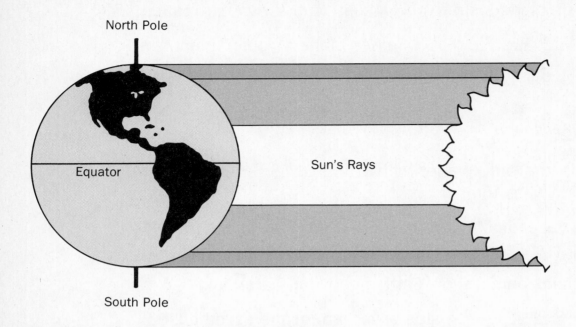

North Pole

Equator

Sun's Rays

South Pole

Earth is the third planet from the sun. It is about 93 million miles from the sun. Earth is a medium-size planet. Mercury, Venus, Mars, and Pluto are smaller the Earth. Jupiter, Saturn, Uranus, and Neptune are much larger than Earth.

Land, water, and air are three major <u>features</u> of Earth. The surface of Earth is made up of soil and rock. Almost three fourths of this surface is covered by water. Around the planet is a layer of air called the <u>atmosphere</u>. Air is a mixture of many different gases.

Earth gets light and heat from the sun. But not all parts of Earth get the same amount of sunlight. The sun shines directly on the equator. There, the rays of the sun are close together. The <u>climate</u> is warm all year long. At the North Pole and the South Pole, the rays of the sun are more spread out. The poles, therefore, are the coldest parts of Earth.

Air, water, and the right temperature make many kinds of plant and animal life possible on Earth. Few living things can live in the cold temperatures near the poles. But many kinds of plants and animals live in the warmer parts of Earth.

A. Write the letter for the correct answer.

1. The surface of Earth is made up of _____c_____.
   (a) water   (b) air   (c) soil and rock

2. Air is a mixture of many different _____b_____.
   (a) rocks   (b) gases   (c) stars

3. Earth gets light and heat from the _____a_____.
   (a) sun   (b) poles   (c) equator

4. The sun shines directly on the part of Earth called the _____c_____.
   (a) North Pole   (b) South Pole   (c) equator

5. Few living things can live in the cold temperatures near the _____c_____.
   (a) soil   (b) equator   (c) poles

B. Answer True or False.

1. Earth is the largest planet. _____False_____

2. Earth is the closest planet to the sun. _____False_____

3. Almost three fourths of Earth's surface is covered by water.
   _____True_____

4. Not all parts of Earth get the same amount of sunlight. _____True_____

5. The rays of the sun are more spread out at the equator. _____False_____

6. Many kinds of plants and animals live in the warmer parts of
   Earth. _____True_____

C. Use each word to write a sentence about Earth.

1. equator _____ Sentences will vary. _____

   _____

2. poles _____

   _____

3. sunlight _____

   _____

13

Earth's Moon

Phases of the Moon

The moon is the largest object in the night sky. It is 240,000 miles from Earth. The moon is our closest neighbor in space.

The moon looks much larger than the stars, but it isn't. Its <u>diameter</u> is only one fourth of Earth's diameter. You may be surprised to learn that the sun is about 400 times larger than the moon.

The moon is a **satellite** of Earth. A satellite is an object in space that orbits another object in space. Earth has one moon. Some planets have many moons.

The moon is the brightest object in the sky at night. But it does not give off light of its own. The moon <u>reflects</u> light from the sun like a giant mirror.

Have you ever seen a full moon? Have you ever seen the moon when it looks like a <u>sliver</u>? During a month, the moon seems to change shape. But it does not really change.

The moon seems to change shape as different amounts of the moon are lit by the sun. As more sunlight hits the moon, more light is reflected to Earth. The amount of sunlight the moon reflects changes a little each night.

The changing views of the moon are called **phases.** The pictures on this page show the phases you can see in a month.

Because the moon and Earth are close, the force of gravity between them is strong. The moon's gravity pulls the waters of Earth toward the moon. This pull makes the level of the water in the oceans change. These changes in water levels are called **tides.**

A. Answer True or False.

1. The moon is the largest object in the night sky. __True__

2. The moon orbits around Earth. __True__

3. The moon is 2,400 miles from Earth. __False__

4. There is no force of gravity between the moon and Earth.
   __False__

5. During a month, we always see the same shape of the moon.
   __False__

6. A satellite is an object in space that orbits another object in space. __True__

B. Fill in the missing words.

1. The changing views of the moon are called __phases__. (phases, tides)

2. Some planets have many __moons__. (orbits, moons)

3. The moon gives off no __light__ of its own. (light, gravity)

4. The moon is a __satellite__ of Earth. (planet, satellite)

5. The changes in the levels of the oceans are called __tides__. (tides, phases)

6. The brightest object in the night sky is the __moon__. (moon, sun)

C. Answer the questions.

1. What causes tides? __The moon's gravity pulls the waters of Earth toward the moon. This pull makes the level of the water in the oceans change.__

2. Why does the moon seem to change shape? __The moon seems to change shape because different amounts of it are lit by the sun at different times.__

The Outer Planets

The five planets farthest from the sun are called the **outer planets.** They are Jupiter, Saturn, Uranus, Neptune, and Pluto. The outer planets are different from Earth. They are very cold and may be made mainly of gas.

Jupiter is the fifth planet in the solar system and the largest. A day on Jupiter is less than 10 hours. But it takes Jupiter 12 Earth years to orbit the sun once. Jupiter is surrounded by thick clouds of gas. Most scientists believe that liquid lies under the clouds.

Saturn is the sixth planet from the sun. It is the second largest planet in the solar system. Saturn is surrounded by shiny rings that make it very colorful. It takes almost 30 Earth years for Saturn to make one orbit around the sun. Saturn has 21 moons.

Uranus is the seventh planet from the sun. Uranus also has rings surrounding it. The rings are formed by an unknown black material. It takes Uranus 84 years to orbit once around the sun.

Scientists discovered Neptune and Pluto by using mathematics. Neptune is the eighth planet from the sun. It takes Neptune about 165 years to orbit the sun one time. Neptune is covered with a thick layer of icy clouds. The interior of Neptune may be a central rocky core.

Pluto is the farthest planet from the sun and the one we know the least about. From Pluto, the sun would probably look like a distant star.

Jupiter

Saturn

A. Circle the letter of the correct answer.

1. What is the largest planet of the solar system?

   (a) Jupiter   (b) Saturn   (c) Pluto

2. Which planet is surrounded by shiny rings?

   (a) Jupiter   **(b) Saturn**   (c) Pluto

3. Which planet is <u>not</u> an outer planet?

   (a) Pluto   (b) Neptune   **(c) Mercury**

4. Which planets did scientists discover by using mathematics?

   (a) Jupiter and Saturn   **(b) Neptune and Pluto**
   (c) Mars and Venus

B. Answer <u>True</u> or <u>False</u>.

1. The outer planets are very hot. ___False___

2. The outer planets may be made mainly of gas. ___True___

3. The rings around Uranus are formed by an unknown black material. ___True___

4. It takes Neptune 30 Earth years to orbit the sun. ___False___

5. The outer planets are very different from Earth. ___True___

6. Saturn is the third planet from the sun. ___False___

7. The planet we know the most about is Pluto. ___False___

C. Write the names of the outer planets. ___Jupiter, Saturn, Uranus, Neptune, Pluto___

D. Answer the questions.

1. In what two ways are the outer planets different from Earth? ___They are very cold and may be made mainly of gas.___

2. How would the sun look from Pluto? ___From Pluto, the sun would probably look like a distant star.___

Exploring Space

An Astronaut in Space

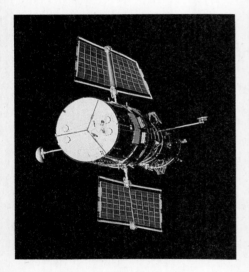
The Hubble Space Telescope

Ancient people did not have tools to look at the stars and the planets. What they learned was based on what they saw with their eyes. Today, people have many tools to study space.

Telescopes are tools that make distant objects look larger. They help people learn about the stars and the planets. With a special kind of telescope, a scientist can learn how far away a star is and what the star is made of.

Satellites also help us explore space. Satellites made by people send pictures and other information back to Earth. The Hubble Space Telescope orbits Earth. It sends us pictures of stars 100 times clearer than we get from instruments on Earth.

Another way people explore space is to send astronauts in **spacecraft.** Astronauts have been to the moon several times. The first landing on the moon took place on July 20, 1969. Neil Armstrong was the first person to walk on the moon. He and other astronauts took many photographs and collected rocks and soil. They left instruments on the moon to gather information.

Astronauts have not traveled to other planets. But some spacecraft have traveled far into space. The Voyager space probes were sent into space to study the most distant planets. Voyager took pictures of Jupiter, Saturn, Uranus, and Neptune. These pictures showed moons that people had never seen before. They showed volcanoes, too. Voyager is now traveling beyond our solar system.

A. Fill in the missing words.

1. Today, people _____ have _____ many tools to study space. (do not have, have)

2. _____ Telescopes _____ are tools that make distant objects look larger. (Telescopes, Eyes)

3. Astronauts have been to _____ the moon _____ several times. (Mars, the moon)

4. _____ Satellites _____ made by people can send photographs and other information back to Earth. (Satellites, Constellations)

5. _____ Voyager _____ is now traveling beyond our solar system. (Voyager, The moon)

B. Answer True or False.

1. Some spacecraft have been to the most distant planets. _____ True _____

2. One kind of special telescope helps scientists tell what a star is made of. _____ True _____

3. Satellites made by people send rocks back to Earth. _____ False _____

4. Astronauts have been to all the planets. _____ False _____

5. Neil Armstrong was the first person to walk on the moon. _____ True _____

6. The pictures from Voyager showed moons that people had not seen before. _____ True _____

C. Draw lines to complete the sentences.

1. Telescopes   took pictures of Jupiter, Saturn, Uranus, and Neptune.

2. Astronauts   send photographs and other information back to Earth.

3. Voyager   are tools that make distant objects look larger.

4. Satellites   have been to the moon several times.

A Year

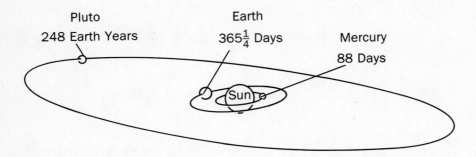

Pluto
248 Earth Years

Earth
$365\frac{1}{4}$ Days

Mercury
88 Days

Sun

A **year** on Earth is 365 days. This is the time it takes for Earth to orbit once around the sun. The exact amount of time is $365\frac{1}{4}$ days. Because of the extra $\frac{1}{4}$ day, 1 day is added to the calendar every 4 years. That year is called leap year. February usually has 28 days, but in a leap year it has 29 days. So there are 366 days in a leap year. All other years have 365 days.

A year is a different length of time on each planet. This is because it takes each planet a different amount of time to orbit the sun. Mercury is much closer to the sun than Earth is. It takes Mercury only 88 days to travel around the sun. So a year on Mercury is 88 days long.

Pluto is very far from the sun. It takes Pluto a long time to travel once around the sun. A year on Pluto is equal to 248 years on Earth.

Fill in the missing words.

1. The time it takes Earth to orbit once around the sun is called a

    ___year___ . (day, year)

2. A year on Earth is ___$365\frac{1}{4}$ days___ . (1 day, $365\frac{1}{4}$ days)

3. In a leap year, February has ___29 days___ . (366 days, 29 days)

4. A year on Pluto is equal to ___248 years on Earth___ .
    (1 year on Earth, 248 years on Earth)

Day and Night

An <u>imaginary</u> line goes through Earth from the North Pole to the South Pole. This line is called Earth's **axis.** Earth spins on its axis. This spinning movement is called **rotation.** Earth rotates once every 24 hours. Each 24 hours is 1 day. Both daytime and nighttime are part of this day.

Only one side of Earth faces the sun at any time. This side is having daytime. The opposite side of Earth is in darkness. When it is daytime in North America, it is nighttime in India. In a 24-hour day, each part of Earth moves into the sunlight and back into darkness again.

The spinning motion of Earth makes the sun look like it is moving across the sky. Because Earth spins toward the east, the sun appears to rise in the east and set in the west. If Earth did not rotate, half of our planet would always have day and half would always have night.

Answer the questions.

1. What is the imaginary line that goes through Earth from the North Pole to the South Pole called? _____ Earth's axis _____

2. Why does it seem that the sun is moving across the sky? _The_ **spinning motion of Earth makes the sun look like it is moving across the sky.**

3. What does Earth do once every 24 hours? _____
    Earth rotates once every 24 hours.

The Seasons

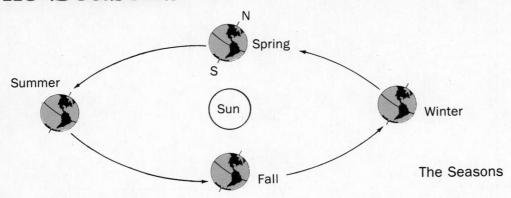

The Seasons

The movement of Earth around the sun and the tilt of Earth are the two reasons why we have seasons. As Earth orbits the sun, the North Pole is tilted toward the sun for part of the year. It is tilted away from the sun for part of the year.

When the North Pole is tilted toward the sun, the northern part of Earth has long days and short nights. It is summer in the **Northern Hemisphere,** the part of Earth above the equator.

When the North Pole is tilted away from the sun, the part of Earth above the equator has short days and long nights. It is winter in the Northern Hemisphere.

There are two seasons when Earth's axis is not tilted toward or away from sun. This is when we have spring and fall. During spring and fall, days and nights have about the same number of hours.

Answer the questions.

1. What causes the seasons? _____ The movement of Earth and the _____ tilt of Earth are the two reasons we have _____ seasons. _____

2. What happens when the North Pole is tilted toward the sun? _____ The northern part of Earth has long days and short nights. _____ It is summer in the Northern Hemisphere. _____

Part A

Use the words below to complete the sentences.

constellation	moon	planets
gas	orbit	stars
gravity	phases	

1. The solar system is made up of the sun and nine _____planets_____.

2. All the planets are held in orbit by a force called _____gravity_____.

3. The curved path of a planet around the sun is called an _____orbit_____.

4. The sun is a large ball of _____gas_____.

5. The _____stars_____ give off light and heat.

6. Ursa Major is an example of a _____constellation_____.

7. A small object in orbit around a planet is a _____moon_____.

8. Different views of the moon are called _____phases_____.

Part B

The names of the planets are listed below. Number them in the order of their distance from the sun. The first one is done for you.

Earth ___3___   Mercury ___1___   Jupiter ___5___

Uranus ___7___   Neptune ___8___   Saturn ___6___

Venus ___2___   Pluto ___9___   Mars ___4___

Part C

Read each sentence. Write <u>True</u> if the sentence is true. Write <u>False</u> if the sentence is false.

1. There would be no life on Earth without the sun. ___True___

2. Planets give off their own light. ___False___

3. Earth rotates, but it does not travel around the sun. ___False___

Model the Seasons

You Need

- a partner
- $\frac{1}{4}$ stick modeling clay
- toothpick
- tennis ball

1. Review the model of the seasons on page 22. To show how Earth moves around the sun, add arrows to the drawing shown here.

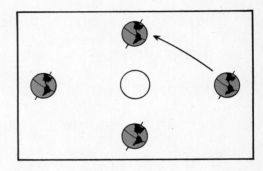

2. Roll the clay into a ball. This will be Earth. Push the toothpick through the center of the ball to make the axis.

3. The tennis ball will be the sun. One partner will hold the sun. The other partner will hold Earth. Remember to tilt Earth as it is in the drawing.

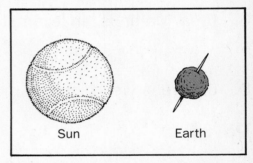

Sun   Earth

4. Spin Earth slowly to show how it rotates. How many times does Earth spin to make one day?

5. Keep spinning Earth, and slowly walk in a circle around the sun. Be sure to keep the axis tilted. How does the tilt of Earth cause winter?

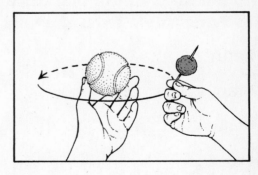

Write the Answer

Describe Earth's tilt when it is winter in the Northern Hemisphere.

When the North Pole is pointed away

from the sun it is winter

in the Northern Hemisphere.

Fill in the circle in front of the word or phrase that best completes each sentence. The first one is done for you.

1. The nine objects in orbit around the sun are
   a. rings.
   ● planets.
   c. telescopes.

2. The only planet that has living things, as we know them, is
   a. Venus.
   b. Jupiter.
   c. Earth.

3. A pattern of stars is called
   a. an asteroid.
   b. a planet.
   c. a constellation.

4. A tool that helps people learn about the stars and the planets is a
   a. telescope.
   b. moon.
   c. equator.

5. The force of gravity between Earth and the moon causes changes in
   a. phases.
   b. tides.
   c. rings.

6. The imaginary line that goes through Earth is its
   a. rotation.
   b. orbit.
   c. axis.

Fill in the missing words.

7. Stars are made of _____hot gases_____. (frozen water, hot gases)

8. A year on Earth is _____365 days_____. (365 days, 24 hours)

9. The sun is a _____star_____. (star, planet)

Write the answer on the lines.

10. What does Earth get from the sun?

    Earth gets heat and light from the sun.
    _____

UNIT 2
Weather

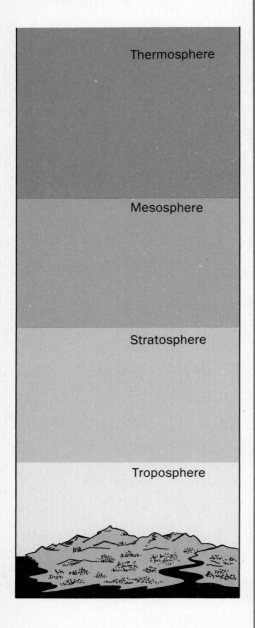

Thermosphere

Mesosphere

Stratosphere

Troposphere

Weather and the Atmosphere

Weather is the condition of the air around us. Look outside. The weather may be hot or cold, wet or dry. Weather is caused by four factors. The factors are temperature, moisture, air pressure, and wind. You will learn about these factors in this unit.

Earth is surrounded by a blanket of air called the atmosphere. All weather develops in the atmosphere. The atmosphere is a mixture of gases. About three fourths of the atmosphere is nitrogen. Almost one fourth is oxygen. The rest of the atmosphere is carbon dioxide and other gases.

These gases are important to living things. People and other animals use oxygen as they breathe. Plants use carbon dioxide to make food. Living things use nitrogen to make proteins.

The atmosphere is in layers around Earth. The layer closest to Earth is the **troposphere**. This layer is 6 to 10 miles thick. Only this layer contains enough air for living things to breathe. Nearly all weather occurs in this layer.

Next is the **stratosphere**. This layer reaches from the troposphere to 30 miles above Earth. The stratosphere contains an important gas called ozone. Ozone keeps most of the sun's harmful radiation from reaching Earth.

Farther out are the **mesosphere** and **thermosphere**. The air in the thermosphere becomes thinner and thinner until it fades into space.

A. Fill in the missing words.

1. Earth is surrounded by a blanket of air called ___the atmosphere___. (the atmosphere, oxygen)

2. The atmosphere is a mixture of ___gases___. (gases, rocks)

3. About three fourths of the atmosphere is ___nitrogen___. (oxygen, nitrogen)

4. The layer of atmosphere closest to Earth is the ___troposphere___. (troposphere, carbon dioxide)

5. Ozone keeps most of the sun's harmful ___radiation___ from reaching Earth. (radiation, sound waves)

B. Answer True or False.

1. Weather is the condition of the air around us. ___True___

2. Only the stratosphere contains enough air for living things to breathe. ___False___

3. All weather occurs in the mesosphere. ___False___

4. The air in the thermosphere becomes thinner and thinner until it fades into space. ___True___

5. There are several layers of the atmosphere. ___True___

6. Nitrogen keeps most of the sun's harmful radiation from reaching Earth. ___False___

C. Answer the questions.

1. What are the four factors that determine weather? ___Weather is caused by temperature, moisture, air pressure, and wind.___

2. What are the layers of gases around Earth? ___The atmosphere is in layers around Earth.___

3. What gas is used by people and other animals as they breathe?
___People and other animals use oxygen as they breathe.___

27

The Water Cycle

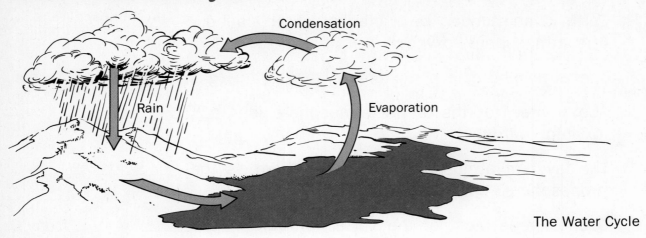

Condensation

Rain

Evaporation

The Water Cycle

The **water cycle** greatly affects weather and living things. You might think that rain brings new water to Earth. But there is no new water on Earth. The water cycle is the movement of water between the ground and the atmosphere.

The first step of the water cycle involves the sun. The sun heats water on Earth and changes it from a liquid to a gas called **water vapor**. The water vapor goes into the air. This process is called **evaporation**. It is evaporation that dries a puddle in the sun after the rain.

In the second step of the cycle, water vapor in the air changes back to a liquid. This happens when warm air holding water vapor gets cooler. This process is called **condensation**. The tiny water drops in the air are packed closer and closer together until they start to join. Clouds are formed when thousands of drops are so close together that you cannot see through them.

As clouds become cooler, the drops join together to make bigger drops. Finally, they become heavy enough to fall back to Earth as rain or snow. This falling rain or snow is the third step in the water cycle.

The word cycle usually means circle. The water cycle is like a circle because it has no beginning or end. There really is no new water on Earth.

A. Answer True or False.

1. There is no new water on Earth. _____True_____

2. Clouds are formed when rain falls. _____False_____

3. The water cycle is the movement of water between the ground and the atmosphere. _____True_____

4. The word cycle usually means square. _____False_____

5. The water cycle greatly affects weather and living things. _____True_____

B. Write the letter for the correct answer.

1. The first step of the water cycle involves the ____a____.
   (a) sun   (b) moon   (c) snow

2. Condensation takes place when warm air holding water vapor gets ____b____.
   (a) warmer   (b) cooler   (c) wet

3. It is ____a____ that dries a puddle in the sun.
   (a) evaporation   (b) rain or snow   (c) atmosphere

4. In the third step of the water cycle, rain or snow ____c____.
   (a) dries up   (b) goes into clouds   (c) falls back to Earth

5. As clouds become cooler, drops join together to make ____b____.
   (a) smaller drops   (b) bigger drops   (c) vapor

C. Answer the questions.

1. What is the water cycle? _____The water cycle is the movement of water between the ground and the atmosphere._____

2. Why is the water cycle like a circle? _____It is like a circle because it has no beginning or end._____

3. When are clouds formed? _____Clouds are formed when thousands and thousands of drops of water are so close together that you cannot see through them._____

Clouds

Cirrus Clouds

Cumulus Clouds

Stratus Clouds

Cumulonimbus Clouds

A cloud is a mass of water droplets or ice crystals that floats in the air. When warm air rises and cools, the water vapor in air condenses and forms water droplets. These droplets form clouds.

There are three main types of clouds. They are **cirrus, cumulus,** and **stratus**. These types of clouds may combine to form many other kinds of clouds. You can study clouds to find out how the weather will change.

Cirrus clouds are white and feathery. They form high in the sky. It is so cold in this part of the atmosphere that cirrus clouds are made entirely of ice crystals. Cirrus clouds usually mean that rain or snow is coming within a day.

Cumulus clouds are thick, white, and fluffy. They look like piles of cotton. You can often see cumulus clouds on a sunny summer day. They mean fair weather. Cumulus clouds form much lower in the atmosphere than cirrus clouds and are made only of water droplets.

Stratus clouds are layers of gray clouds that cover most of the sky. They often mean that rain or snow is coming. Stratus clouds form low in the atmosphere. A stratus cloud that forms on the ground is called **fog**. Fog forms when air that is holding a lot of water cools off quickly at night.

Sometimes the term nimbus is added to the name of a cloud. Nimbus means that a cloud is heavy and dark and will bring rain. Cumulonimbus clouds, for example, produce summer thunderstorms.

A. **Draw lines to match the clouds with their descriptions.**

1. cirrus —————— white and feathery

2. cumulus ⟍ ⟋ layers of gray clouds that cover most of the sky

3. stratus ⟋ ⟍ piles of cotton

B. **Fill in the missing words.**

1. Cirrus clouds form ___high___ in the sky. (high, low)

2. You can often see cumulus clouds on a ___sunny___ summer day. (sunny, rainy)

3. Cirrus clouds usually mean that _____rain or snow_____ is coming within a day. (rain or snow, a thunderstorm)

4. A stratus cloud that forms on the ground is called ___fog___. (cirrus, fog)

5. ___Nimbus___ means that a cloud is heavy and dark and will bring rain. (Nimbus, Cumulus)

6. Cumulonimbus clouds produce summer _____thunderstorms_____. (sunny days, thunderstorms)

7. You can study clouds to find out how the ___weather___ will change. (weather, moon)

C. **Write the names of each type of cloud on the lines below.**

___cirrus___   ___cumulus___   ___stratus___

D. **Answer the questions.**

1. What is a cloud? ____A cloud is a mass of water droplets____
___or ice crystals that floats in the air.___

2. What are cirrus clouds made of? ____Cirrus clouds are____
___made entirely of ice crystals.___

Precipitation

Rain

Snowflakes

Hail

When water falls to Earth, it is called **precipitation**. Rain, snow, sleet, and hail are all forms of precipitation.

For rain to fall, the water drops in a cloud must reach a certain size. As the tiny droplets of water in a cloud bump into each other, they combine. When the drops are heavy enough, they fall as rain.

If the air is very cold, water vapor in a cloud may freeze into tiny ice crystals called snow. All snowflakes have six sides, or points.

If rain falls through a layer of very cold air before reaching the ground, the drops will freeze. They form pieces of ice called sleet. Sleet forms only in cold weather.

Hail may fall during a thunderstorm in the spring or summer. Hail begins as tiny pieces of ice in a thundercloud. The pieces of ice start to fall, but they are swept back up by a current of air. As the pieces fall again, they collect more water. This water freezes, making the hail larger. This happens over and over again until the hail becomes so heavy it falls to the ground.

Sometimes you will find that blades of grass have drops of water on them even when it has not rained. This water is called **dew**. Dew forms when there is a large amount of water vapor in the air. At night the air cools off. Cool air cannot hold as much moisture as warm air, so the water vapor condenses into drops of dew. If the ground is very cold, the dew freezes and forms **frost**.

A. Fill in the missing words.

1. For rain to fall, the water drops in a cloud must reach a certain ___size___. (size, color)

2. As the tiny droplets of water in a cloud bump into each other, they ___combine___. (get smaller, combine)

3. Sleet forms only in ___cold weather___. (hot weather, cold weather)

4. All ___snowflakes___ have six sides, or points. (snowflakes, hail)

5. When dew freezes, it forms ___frost___. (frost, rain)

B. The steps below tell how hail is formed. Number the steps in the correct order. The first one is done for you.

___1___ Hail begins as tiny pieces of ice.

___5___ The hail is so heavy it falls to the ground.

___2___ The pieces of ice start to fall, but are swept back up by a current of air.

___4___ This water freezes, making the hail larger.

___3___ As the pieces fall again, they collect more water.

C. Answer True or False.

1. When water drops are heavy enough, they freeze. ___False___

2. Sleet is formed when rain falls through a layer of very cold air and freezes. ___True___

3. Hail may fall during a thunderstorm in spring or summer. ___True___

4. Dew forms when there is a large amount of water vapor in the air. ___True___

D. Answer the question.

What must happen before rain can fall? ___Before rain can fall, the water drops in a cloud must reach a certain size.___

33

Humidity

Polar regions have low humidity because of their cold temperatures.

Regions near the equator usually have high humidity.

The amount of water vapor in the air is called **humidity**. The more moisture there is in the air, the higher the humidity. The humidity is not the same every day. It changes depending on how warm or cold the air is. Warm air can hold more water vapor than cold air.

Sometimes on warm days, the air feels moist. There is so much water vapor in the air that the air cannot hold any more. On these days, the humidity is very high.

You feel cool and comfortable when perspiration from your body can evaporate into the air. But on humid days, the air cannot hold much more water. So perspiration does not evaporate easily. You feel hot, sticky, and uncomfortable.

Some places tend to be more humid than others. Coastal regions often have high humidity because water from the ocean evaporates, putting more water vapor into the air. Deserts usually have low humidity because there is little water to evaporate into the air.

Warm air holds more water vapor than cold air. That is why places near the equator are more humid than the poles.

Answer <u>True</u> or <u>False</u>.

1. The amount of water vapor in the air is called humidity. __True__

2. The more moisture there is in the air, the lower the humidity.
   __False__

3. Deserts usually have low humidity. __True__

4. Warm air can hold more water vapor than cold air. __True__

Temperature

Temperature is a measure of how hot or cold something is. The temperature of the air greatly affects the weather. Energy from the sun travels through the atmosphere to Earth. About half of the sun's energy hits the surface of Earth. Part of the sun's energy is reflected by clouds and dust in the air. Some of the energy warms Earth's surface. Heat from Earth's surface warms the atmosphere around it.

Temperature is measured in **degrees** using a <u>thermometer</u>. The higher the number of degrees, the warmer the atmosphere.

Two scales of degrees are used on thermometers to measure temperature. One is the <u>Celsius</u> scale. The other is the <u>Fahrenheit</u> scale. On a very hot day, the temperature may reach close to 100 degrees Fahrenheit. This would be about 38 degrees Celsius. Water freezes at 32 degrees Fahrenheit. This would be 0 degrees Celsius. Average room temperature is 72 degrees Fahrenheit. In place of the word degree, you can draw a small circle. So 72 degrees can also be written as 72°.

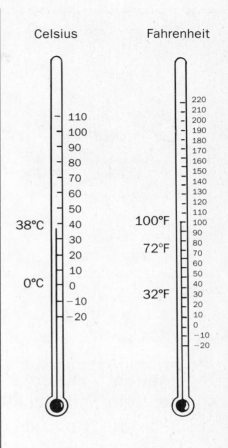

Thermometers

Fill in the missing words.

1. ___Temperature___ is a measure of how hot or cold something is. (Temperature, Humidity)

2. On a very hot day, the temperature may be ___38°___ Celsius. (38°, 0°)

3. On the Fahrenheit scale, water freezes at ___32°___. (212°, 32°)

4. Heat from Earth's ___surface___ warms the atmosphere. (surface, clouds)

What Makes the Weather Change?

Warm Front

Warm Air

Cold Air

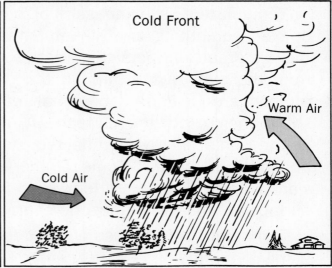

Cold Front

Warm Air

Cold Air

The kind of weather you are having right now depends on the kind of **air mass** that is over your area. An air mass is a huge body of air that stays together as it moves. It has about the same temperature and humidity all through it.

There are four main kinds of air masses. They are cold and dry, cold and wet, warm and dry, and warm and wet. Air masses can form over land or water. Those that form over land have dry air. Those that form over water have wet air. When an air mass forms near the equator, it has warm air. When it forms near the poles, it has cold air.

When two air masses meet, they do not mix. Instead, they form a boundary, or **front**. Most changes in the weather take place along fronts.

A cold front forms when a cold air mass moves into a warm air mass. The weather changes suddenly when a cold front arrives. Cold fronts may bring partly cloudy skies, rain, or snow. Cooler temperatures and clear skies often follow the front.

A warm front forms when a warm air mass moves in over a cold air mass. Warm fronts may bring wispy clouds, light rain, or snow. The weather that follows is warm and humid.

A. Fill in the missing words.

1. The kind of weather you are having right now depends on the kind of ___air mass___ that is over your area. (air mass, sleet)

2. An air mass has about the same ___temperature___ and humidity all through it. (temperature, clouds)

3. Air masses that form over land have ___dry___ air. (wet, dry)

4. When two air masses meet, they form a ___front___. (front, pole)

5. Most changes in the weather take place along ___fronts___. (poles, fronts)

6. When an air mass forms near the poles, it has ___cold___ air. (cold, warm)

B. Name the four kinds of air masses.

cold and dry	warm and dry
cold and wet	warm and wet

C. Answer True or False.

1. An air mass is a huge body of air that stays together. ___True___

2. There are 12 main kinds of air masses. ___False___

3. An air mass that forms near the equator has cold air. ___False___

4. When air masses meet, they mix. ___False___

5. The weather changes suddenly when a cold front arrives. ___True___

6. Cold fronts may bring partly cloudy skies, rain, or snow. ___True___

D. Answer the questions.

1. What kind of weather follows a cold front? ___Cooler temperatures and clear skies often follow a cold front.___

2. What kind of weather follows a warm front? ___The weather that follows a warm front is warm and humid.___

Air Pressure

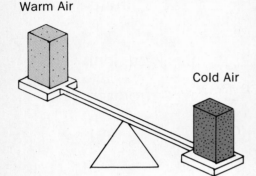

Warm Air

Cold Air

Cold air is heavier than warm air.

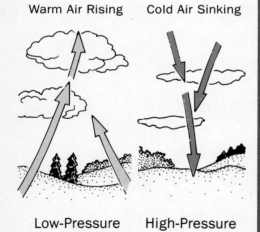

Warm Air Rising   Cold Air Sinking

Low-Pressure Area   High-Pressure Area

One factor that greatly affects the weather is **air pressure**. Air is made of many gases. These gases are made of tiny particles. The particles have weight and push down on Earth. The push of air on Earth is called air pressure.

Air pressure is changed by temperature. The top picture shows how temperature changes the pressure of air. When air is heated, the particles of air move apart. When air is cooled, the particles move closer together. Notice that the blocks of air are not balanced. The cold air has more particles of air than the warm air. The cold air is heavier and has more pressure.

Air pressure is different in warm and cold places. The bottom picture compares two places on Earth. The particles of air above the warm place begin to spread out. The warm air becomes light and starts to rise. As warm air rises, it has less pressure. Rising air forms a low-pressure area, called a **low**. Cold air is heavier than warm air because the particles are packed tightly together. Since cold air is heavier than warm air, it falls to the ground. Falling cold air forms a high-pressure area, called a **high**.

The movement of air from a high-pressure area to a low-pressure area causes **wind**. The greater the difference in pressure between the areas, the stronger the wind will be.

People who study weather measure the air pressure every day. Air pressure is measured with a **barometer**.

A. Fill in the missing words.

1. Air is made of many ____gases____. (liquids, gases)

2. The push of air on Earth is called ____air pressure____.
   (humidity, air pressure)

3. Air pressure is changed by ____temperature____.
   (temperature, rain)

4. When air is heated, the particles of air move ____apart____.
   (closer together, apart)

5. The greater the difference in pressure between areas, the
   ____stronger____ the wind will be. (stronger, weaker)

6. When air is cooled, the particles move ____closer together____.
   (farther apart, closer together)

B. Answer True or False.

1. Cold air is heavier than warm air. ____True____

2. As warm air rises, it has more pressure. ____False____

3. Air pressure is the same in all places. ____False____

4. Falling cold air forms a high-pressure area, called a high.
   ____True____

5. Air pressure is measured with a barometer. ____True____

6. Cold air and warm air have the same air pressure. ____False____

C. Answer the questions.

1. What causes wind? ____The movement of air from a high-pressure
   area to a low-pressure area causes wind.____

2. What is a low? ____Rising air forms a low-pressure
   area, called a low.____

3. What is a high? ____Falling cold air forms a high-pressure
   area, called a high.____

Storms

Tornado

Changes in weather take place when air masses meet. For example, warm air masses meeting cold air masses can cause rain. If there is a large difference in the temperature of the air masses, then storms may result.

Thunderstorms take place mainly in the spring and summer in North America. There is heavy rain during a thunderstorm. There is **thunder** and **lightning**, too. Lightning is caused by electric charges in a cloud. These charges can jump from one cloud to another. Or they can jump from a cloud to the ground. As it passes through the air, lightning heats up the air and makes it expand very quickly. The noise that the expanding air makes is thunder.

A **tornado** is a <u>violent</u> but small storm that starts over land. In a tornado, winds begin to move in a circle. They may go as fast as 500 miles an hour. The winds make a tall, dark cloud in the shape of a <u>funnel</u>. The bottom of the funnel may touch the ground. If it does, a great deal of damage can occur. Buildings are destroyed. Cars and other objects are lifted and thrown many yards away.

A **hurricane** is a large storm that forms over an ocean near the equator. It may be 500 miles across and have winds of 75 to 200 miles an hour. The winds spin around a center, called the <u>eye</u>. Inside the eye of a hurricane, it is calm. There is no wind. Since hurricanes form over oceans, they often damage islands and coastal areas with their high winds and heavy rains.

A. Fill in the missing words.

1. If there is a large difference in the temperature of air masses that meet, _____storms_____ may result. (fair weather, storms)

2. Thunderstorms take place mainly in spring and __summer__ in North America. (winter, summer)

3. Lightning is caused by _____electric charges_____ in a cloud. (wind, electric charges)

4. A tornado starts over __land__. (land, water)

B. Draw lines to match the kinds of storms with their descriptions.

1. thunderstorms ——————— heavy rain, with thunder and lightning

2. tornado — large storm that forms over an ocean

3. hurricane — a violent but small storm that starts over land

C. Answer True or False.

1. There is heavy rain during a thunderstorm. __True__

2. In lightning, electric charges can move from a cloud to the ground. __True__

3. The noise that expanding air makes is thunder. __True__

4. A hurricane is shaped like a funnel. __False__

5. The center of a hurricane is called the mouth. __False__

D. Answer the questions.

1. What kind of damage occurs when the funnel of a tornado touches the ground? __Buildings are destroyed. Cars and other__ __objects are lifted and thrown many yards__ __away.__

2. What causes the damage in a hurricane? __The high winds__ __and heavy rains cause damage.__

41

Tracking the Weather

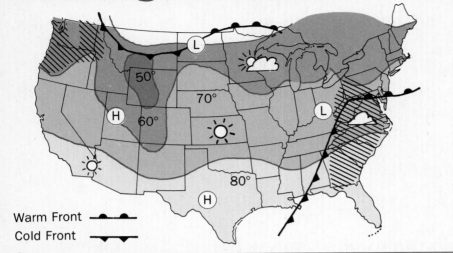

H High Pressure
L Low Pressure
▨ Rain
☼ Sunny
⛅ Partly Cloudy
☁ Cloudy

Warm Front ━●━●━
Cold Front ━▼━▼━

People who study weather are **meteorologists**. They measure air pressure, temperature, humidity, wind speed, precipitation, and cloud types. They try to predict how the weather will change.

Meteorologists have many instruments to help them. Weather vanes tell the direction of the wind. Weather balloons measure the conditions of the atmosphere. Weather satellites are sent up into space. Information that they send back helps meteorologists track hurricanes and other storms.

Meteorologists use the information they have collected to make weather maps. These maps show areas of high and low pressure. They show warm and cold fronts. Look at the weather map on this page. Follow one of the bands that crosses the country. This band has the same temperature at the beginning and at the end. It shows the parts of the country that have the same temperature.

Answer True or False.

1. Meteorologists study fossils. _____False_____

2. Weather maps show areas of high and low pressure. _____True_____

3. Weather vanes measure humidity. _____False_____

4. Meteorologists have many instruments to help them. _____True_____

Part A

Use the words below to complete the sentences.

atmosphere	humidity	troposphere
evaporation	hurricane	water cycle
front	stratus clouds	weather

1. Earth is surrounded by a blanket of air called the _____ atmosphere _____ .

2. The condition of the air around you is called _____ weather _____ .

3. The layer of the atmosphere closest to Earth is the _____ troposphere _____ .

4. The movement of water between the ground and the atmosphere is the _____ water cycle _____ .

5. Water vapor goes into the air by a process called _____ evaporation _____ .

6. Layers of gray clouds that cover the sky are _____ stratus clouds _____ .

7. The amount of water vapor in the air is called _____ humidity _____ .

8. When two air masses meet, they form a _____ front _____ .

Part B

Read each sentence. Write <u>True</u> if the sentence is true. Write <u>False</u> if the sentence is false.

1. A cloud is a mass of water droplets or ice crystals. ___True___

2. The water in clouds falls back to Earth as rain or snow. ___True___

3. Rain brings new water to Earth. ___False___

4. Temperature is measured in degrees. ___True___

5. The more moisture there is in the air, the higher the humidity. ___True___

6. People who study weather are meteorologists. ___True___

7. Lightning is caused by electric charges in a cloud. ___True___

Record Temperatures

You Need

- ● a partner
- ● 4 thermometers
- ● paper
- ● pencil

1. Choose four places inside the classroom to place thermometers. Include a high spot, low spot, dark corner, and one place near a window or a door.

2. Draw a chart. Label it with the headings **Place, Fahrenheit Temperature,** and **Celsius Temperature.** Write on the chart where you placed each thermometer.

3. Check each thermometer in one hour.

4. Write both the Fahrenheit and Celsius temperatures on your chart. Is either the Fahrenheit or Celsius temperature always higher? If so, which one?

5. Compare the temperatures of the different places. Which spot has the highest temperature? Which spot has the lowest temperature? Why?

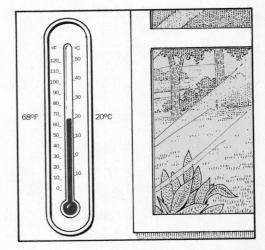

Place	Fahrenheit Temperature	Celsius Temperature
on cabinet		
by window	68°	20°
in closet		
under desk		

Write the Answer

Explain why the two spots with the middle temperatures are warmer and cooler than the other two spots.

Answers will vary but should include the concept that areas exposed to

or sheltered from direct heat sources will register temperatures at the extremes.

Other areas will register temperatures in between the extremes.

Fill in the circle in front of the word or phrase that best completes each sentence. The first one is done for you.

1. The movement of water from the ground to the air and back again is
   ● the water cycle.
   ⓑ evaporation.
   ⓒ condensation.

2. The push of air on Earth is called
   ⓐ barometer.
   ⬤ⓑ air pressure.
   ⓒ wind.

3. When raindrops freeze they form pieces of ice called
   ⓐ dew.
   ⓑ humidity.
   ⬤ⓒ sleet.

4. The amount of water vapor in the air is called
   ⬤ⓐ humidity.
   ⓑ snow.
   ⓒ hail.

5. Lightning is caused by
   ⓐ tornadoes.
   ⓑ fog.
   ⬤ⓒ electric charges.

6. Most changes in weather take place along
   ⓐ oceans.
   ⬤ⓑ fronts.
   ⓒ poles.

Fill in the missing words.

7. Temperature measures _____heat_____. (heat, rainfall)

8. The atmosphere is a mixture of _____gases_____. (gases, liquids)

9. Clouds are made up of tiny drops of _____water_____. (oxygen, water)

Write the answer on the lines.

10. What can happen if there is a large temperature difference between two nearby air masses?

A large temperature difference between two nearby air masses

can produce storms.

UNIT 3
Land and Water

Barges on the Mississippi

Mississippi River Delta

Rivers

A **river** is a natural flow of water that runs into a lake, ocean, or other body of water. The water always flows from high ground to lower ground. The place where a river starts is called its **source.** The source may be a spring or the snow or ice on a mountain. The bottom of a river is called the **bed** and the sides are called the **banks.** The **mouth** is the end of a river. It is the place where the river empties into another body of water.

Rainfall adds water to a river as it flows from its source to its mouth. When it rains, some water is soaked up by the soil. When the soil cannot soak up any more water, rain runs across the ground. It runs in small paths until it joins larger paths of water. This water is called **run-off.**

The water flowing in rivers is used in many ways. The water in a fast-flowing river can produce electricity. Ships on slower-moving rivers deliver food and other important materials. Many cities located along rivers get all their drinking water from rivers.

The Mississippi River is perhaps the most famous river in North America. This river is nearly 2,400 miles long. It flows from its source in northwestern Minnesota into the Gulf of Mexico. The Missouri River flows into the Mississippi River. A river that flows into another river is called a **tributary.** Long rivers usually have many tributaries. The Nile River in Africa is the longest river in the world. It is over 4,000 miles long.

A. Answer True or False.

1. All rain is soaked up by the soil. _____False_____

2. The place where a river starts is its source. _____True_____

3. The bottom of a river is called the bank. _____False_____

4. The sides of a river are called the beds. _____False_____

5. The mouth is the end of a river. _____True_____

6. Rain that runs across soaked soil is called run-off. _____True_____

7. In a river, water always flows from high ground to lower ground.
   _____True_____

8. The water that flows in a river cannot be used. _____False_____ .

B. Fill in the missing words.

1. A river that flows into another river is called a _____tributary_____ .
   (mouth, tributary)

2. The _____source_____ of a river may be a spring or the snow or ice on
   a mountain. (source, mouth)

3. The water in a fast-flowing river can produce _____electricity_____ .
   (run-off, electricity)

4. Ships on a _____slower-moving_____ river deliver food.
   (fast-flowing, slower-moving)

5. _____Rainfall_____ adds water to a river as it flows from its source to
   its mouth. (Rainfall, Soil)

C. The sentences below tell how rainfall affects rivers. Put the sentences in the correct order. The first one is done for you.

____2____ When the soil cannot soak up any more water, rain runs across the ground.

____1____ When it rains, some water is soaked up by the soil.

____3____ The rain runs across the ground in small paths until it joins larger paths of water.

Lakes and Ponds

A lake can be created when a dam is built on a river.

Lakes and **ponds** are bodies of water that have land all around them. They usually contain **fresh water.** The low ground that holds a lake is called a **basin.** A basin may be almost any size and shape. Lake Erie in North America has a huge basin. It covers over 10,000 square miles. Ponds are often small and shallow enough that sunlight reaches the bottom. Lakes are bigger and deeper than ponds.

Many different plants and animals live in lakes and ponds. Tiny plants grow below the surface of the water in most lakes. Some plants float freely, while others are attached to the bottom of the lake. Larger plants such as lilies grow on the surface of some lakes. Insects, fish, turtles, frogs, and ducks feed on the plants that live in the water.

Ponds and lakes are formed in many ways. Some lakes were made thousands of years ago by large sheets of ice. As these sheets of ice moved, they carved holes in the land. When the ice melted, the water filled the holes, making lakes. Other lakes were formed when rainfall and melted ice filled <u>volcano craters</u> with water. Today, some lakes and ponds are created when people build dams on rivers.

People use ponds and lakes for many things. Lakes may be used as a water supply for cities. Ponds provide water for farms. People use ponds and lakes for fishing, swimming, and boating.

A. Answer True or False.

1. The low ground that holds a lake is called a basin. __True__

2. All lakes and ponds were made long ago. __False__

3. Many different plants and animals live in lakes and ponds.
   __True__

4. People do not use lakes as a water supply. __False__

5. Lakes and ponds usually contain fresh water. __True__

6. Frogs and ducks live near ponds and lakes. __True__

7. Ponds are bigger and deeper than lakes. __False__

B. Use the words below to complete the sentences.

basin	dams	ponds
craters	fishing	water

1. The low ground that holds a lake is called a __basin__.

2. People use lakes and ponds for __fishing__, swimming, and boating.

3. Some lakes were formed when rain filled volcano __craters__.

4. Some lakes and ponds are created when people build __dams__.

5. Ponds provide __water__ for farms.

C. Answer the questions.

1. What is the difference between lakes and ponds? __Lakes are__
   __bigger and deeper than ponds.__

2. What are two ways lakes and ponds are used? __Answers may__
   __vary, but should include: as a water supply__
   __for cities, to provide water for farms,__
   __and for fishing, swimming, and boating.__

49

Wells

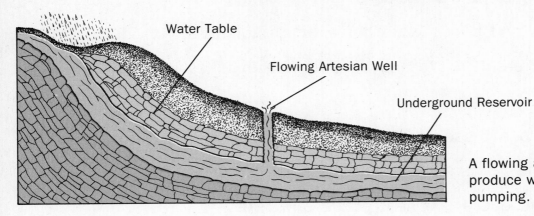

Water Table

Flowing Artesian Well

Underground Reservoir

A flowing artesian well can produce water without pumping.

Sometimes people cannot get all the water they need from nearby rivers, streams, and lakes. Then they may have to look for water under the ground. Underground water is called **ground water.** Ground water is rain that has soaked into the soil. The rainwater moves deeper into the soil until it is stopped by a layer of underground rock.

Ground water usually collects above a layer of rock, forming a reservoir. It may flow in an underground stream. The top of this stream or reservoir is called the **water table.** After a heavy rain or after snow melts, the water table is high. After dry weather, the water table is lower.

To use ground water, **wells** are dug or drilled. A well must be deep enough to reach the water table. Water is pumped up through the well. Many cities and towns use wells to get drinking water. The wells must be watched carefully. If water is pumped out too quickly, the water table will get low. Then deeper wells may have to be dug. Deeper wells may be 100 to 1,000 feet deep.

In some places, underground water flows in a stream, from high areas to lower levels. Pressure builds up at the lower level. A well drilled at the low point will produce water without pumping. This type of well is called a **flowing artesian well.** These wells are usually about 50 feet deep.

A. **Write the letter for the correct answer.**

1. Underground water is called ___b___.
   (a) rain   (b) ground water   (c) soil

2. The top of an underground stream or reservoir is called ___a___.
   (a) the water table   (b) snow   (c) ground water

3. Many cities and towns use ___c___ to get drinking water.
   (a) drills   (b) seas   (c) wells

4. If water is pumped out of the ground too quickly, the water table will get ___c___.
   (a) high   (b) salty   (c) low

5. Water flows freely to the top in ___c___ wells.
   (a) drilled   (b) dug   (c) flowing artesian

6. A well must be dug deep enough to reach the ___a___.
   (a) water table   (b) ocean   (c) lake

B. **Answer True or False.**

1. People can always get all the water they need from nearby rivers, streams, and lakes. ___False___

2. Ground water collects in lakes. ___False___

3. In some places, underground water flows in a stream, from high areas to lower levels. ___True___

4. Deeper wells may be 100 feet to 1,000 feet deep. ___True___

5. Only a few cities and towns use wells. ___False___

6. Flowing artesian wells are usually about 50 feet deep. ___True___

C. **Use each word to write a sentence about wells.**

1. ground water _____ Sentences will vary. _____

   _____

2. water table _____

   _____

Caves

Carlsbad Caverns, New Mexico

Caves are underground holes big enough for a person to enter. Some caves are one small room, and others contain many rooms of different sizes. Larger caves may have underground lakes, rivers, and waterfalls. Mammoth Cave in Kentucky is made up of more than 190 miles of caves. It is the largest cave system ever explored.

Caves are formed by ground water that has dripped or flowed over underground rock for a long time. Slowly the water wore through the rock, leaving caves. The rock found in most caves is **limestone.** A chemical in ground water called carbonic acid breaks down limestone.

In some caves, long pointed rocks hang from the roofs. These rocks are called **stalactites.** They are made by water that drips from the roof. Water containing carbonic acid dissolves part of the limestone. The mineral calcite is left behind. This calcite gathers little by little until a stalactite is formed. With each new drip, more calcite is deposited. The long pointed stalactite gets longer.

Some drops of water fall from the roof of the cave to the floor. Each drop dries, leaving a mineral deposit. Calcite builds up to form long pointed rocks, called **stalagmites,** that rise from the floor.

If a stalactite and a stalagmite meet, they form a tall, thin **column** of rock. Carlsbad Caverns in New Mexico has some of the biggest and most beautiful stalactites, stalagmites, and cave columns in the world.

A. Fill in the missing words.

1. Caves are holes that have been made by _____ground water_____.
   (people, ground water)

2. The rock found in most caves is _____limestone_____.
   (limestone, acid)

3. Some caves contain many ___rooms___. (wells, rooms)

4. In some caves, long pointed rocks hang from the ___roof___.
   (roof, sides)

5. Stalactites are made by water that ___drips___ from the roof of a
   cave. (flows, drips)

6. Stalagmites are rocks that rise from the ___floor___ of a cave.
   (roof, floor)

7. Larger _____caves_____ may have underground lakes, rivers, and
   waterfalls. (limestone, caves)

B. Draw lines to match the words with their descriptions.

1. limestone   large underground holes

2. caves   rock found in most caves

3. carbonic acid —————a chemical in ground water

4. stalactites   water that carved out caves

5. stalagmites   rocks hanging from cave roofs

6. ground water   rocks that rise from cave floors

C. The sentences below tell how stalactites are formed. Put the sentences in the correct order. The first one is done for you.

___3___ A mineral called calcite is left behind.

___2___ The water dissolves part of the limestone.

___1___ Water drips from the roof.

___4___ The calcite gathers little by little until a stalactite is formed.

The Oceans

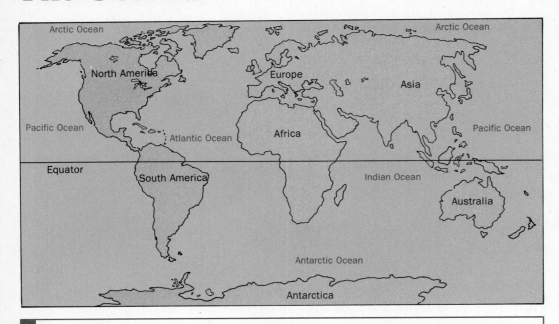

Suppose you are aboard a spacecraft miles above Earth. As you look back at Earth, you can see that almost three fourths of it is covered with water.

Most of the water on Earth is in the **oceans.** The oceans are very large and deep. Some oceans are more than 2 miles deep. The Mariana Trench in the Pacific Ocean is the lowest place on Earth. Here the ocean is almost 7 miles deep! Sunlight cannot reach the deeper layers of the oceans.

The water in the oceans is **salt water.** It is made of many minerals dissolved in water. In fact, all the minerals found on land are also found in salt water. Even gold and silver can be found in salt water. Table salt is the mineral found in the greatest quantity. Ocean water can only be used for drinking if the salt is taken out.

The Pacific Ocean is the largest ocean. It is west of North America. East of North America is the Atlantic Ocean, which is known for its storms.

The Arctic Ocean lies near the North Pole. The Antarctic Ocean lies near the South Pole. The temperature in these oceans may be 28°F, or below freezing. The temperature of the Indian Ocean, near the equator, may reach 85°F.

A. Answer True or False.

1. Almost one third of Earth is covered by water. _____False_____

2. Some oceans are more than 2 miles deep. _____True_____

3. Sunlight can reach all the layers of the oceans. _____False_____

4. The water in the oceans is fresh water. _____False_____

5. The Pacific Ocean is the largest ocean. _____True_____

6. The temperature of ocean water is the same in all parts of the world. _____False_____

B. Fill in the missing words.

1. Most of the water on Earth is in the _____oceans_____. (oceans, clouds)

2. The temperature of the Indian Ocean, near the _____equator_____, may reach 85°F. (equator, poles)

3. Ocean water can only be used for drinking if the _____salt_____ is taken out. (salt, oxygen)

4. Even gold and silver can be found in _____salt_____ water. (fresh, salt)

5. Sunlight cannot reach the _____deeper_____ layers of ocean water. (top, deeper)

6. Table salt is the _____mineral_____ found in the greatest quantity in oceans. (mineral, gold)

C. Answer the questions.

1. Why are the oceans not used much for drinking water? _____Ocean water
can only be used for drinking if the salt is taken out._____

2. What is the temperature of the oceans near the poles? _____The
temperature in these oceans may be 28°F, or
below freezing._____

55

Along the Ocean Floor

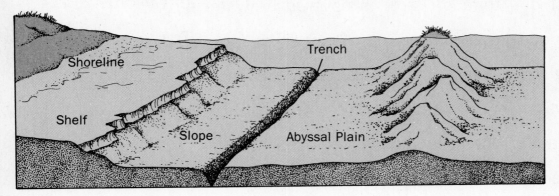

The Ocean Floor

Like dry land, the ocean floor has many features. It is not flat and smooth everywhere. The ocean floor has mountains, valleys, trenches, plains, and even volcanoes!

Imagine being able to walk across the bottom of the Atlantic Ocean. You could start walking at the **shoreline,** where the land and ocean meet. For the first 100 miles, the ocean floor would slowly get deeper. This part of the ocean floor is called the **shelf.** It is mainly smooth and flat. But if you kept walking, you would find yourself walking down a steep hill. This part of the ocean floor is called the **slope.** It lasts for 2 or 3 miles.

At the end of the slope is a wide, flat floor, which is called the **abyssal plain.** It extends for many miles and makes up most of the ocean floor.

If you crossed the abyssal plain, you would climb high mountains and cross deep valleys. You might fall into <u>trenches</u>, deep cuts in the ocean floor.

The mountains below the ocean surface are huge. On the floor of the Atlantic Ocean, there is a mountain range that is 10,000 miles long. It is so high that the peaks stick out of the water. The tops of underwater mountains that show above the surface of the oceans form islands. In the Pacific Ocean, the islands of <u>Hawaii</u> were formed by underwater volcanoes.

A. Fill in the missing words.

1. The ___floor___ of the ocean is not flat and smooth everywhere. (floor, shelf)

2. The ___shoreline___ is where the dry land and ocean meet. (shoreline, slope)

3. The first 100 miles from the shoreline is called the ___shelf___. (shelf, abyssal plain)

4. The ___slope___ is a steep hill that lasts for 2 or 3 miles. (slope, shelf)

5. The abyssal plain is a wide, ___flat___ floor. (hilly, flat)

6. On the floor of the Atlantic Ocean, there is a mountain range that is ___10,000___ miles long. (2,000, 10,000)

7. The tops of underwater mountains that show above the surface of the oceans form ___islands___. (islands, trenches)

8. If you crossed the ocean floor, you would climb high mountains and cross deep ___valleys___. (valleys, plains)

B. Use the words below to complete the sentences.

| features | mountain range | trenches |
| floor | slope | volcanoes |

1. Deep cuts in the ocean floor are called ___trenches___.

2. In the Pacific Ocean, the islands of Hawaii were formed by underwater ___volcanoes___.

3. Some ___features___ of the ocean floor are mountains, valleys, trenches, plains, and even volcanoes.

4. The abyssal plain makes up most of the ocean ___floor___.

5. In the Atlantic Ocean, there is a ___mountain range___ that is 10,000 miles long.

Life in the Oceans

Plants and Animals of the Ocean

The oceans hold some of the largest and smallest animals on Earth. Blue whales are more than 95 feet long. But the smallest sea animals can only be seen with a microscope.

Plants live in the oceans, too. Plants need sunlight and minerals to live. Ocean plants get minerals from ocean water and grow where there is sunlight.

The plants and animals in the ocean can be divided into three groups. The first are those, like jellyfish and plankton, that drift and float in the top layer of ocean water. Plankton are tiny plants and animals that are food for many sea animals.

The second group is made up of thousands of animals that swim freely in the ocean. These animals live below the ocean surface but not on the ocean floor. Octopus, squid, and many kinds of fish belong to this group. Fish are important because they are food for so many people and other animals.

The third group is the plants and animals that live on the floor of the ocean. Plants can only live on the ocean floor near the shore. They cannot live in deep water because they need sunlight.

But many animals can live on the ocean floor. Some of these are worms, snails, starfish, crabs, and lobster. They eat dead animals that fall from above and help to keep the ocean floor clean.

A. Answer True or False.

1. The oceans hold some of the largest and smallest animals on Earth. ___True___

2. Animals can only live on the ocean floor near the shore because they need sunlight. ___False___

3. Some plants and animals drift and float in the top layer of ocean water. ___True___

4. Worms, snails, starfish, crabs, and lobster can live on the ocean floor. ___True___

5. Ocean plants get minerals from ocean water. ___True___

6. There are very few animals that swim freely in the ocean. ___False___

7. The plants and animals that live in the ocean can be divided into three groups. ___True___

B. Draw lines to match each living thing with its description.

1. blue whale   tiny plants and animals
2. plankton   food for many people and animals
3. fish   live on the floor of the ocean
4. crabs   more than 95 feet long

C. Answer the questions.

1. What are plankton? ___Plankton are tiny plants and animals that are food for many animals.___

2. Why are fish important? ___Fish are important because they are food for so many people and other animals.___

Continents

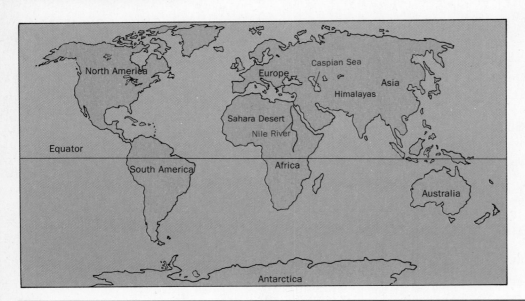

Continents are the largest land masses. The seven continents are **North America, South America, Europe, Asia, Africa, Australia,** and **Antarctica.**

North America is the continent with the longest coastline. It has many rivers, mountain ranges, and plains. Many kinds of weather are found in North America. North America is connected to the continent of South America. Much of South America has hot weather and is known for its rain forests. But the continent's tip is close to the cold South Pole.

Europe, Asia, and Africa are joined together. Europe has a jagged coastline and mild weather. It has many rivers, lakes, and high mountains. The biggest lake in the world lies between Europe and Asia. It is the Caspian Sea. Asia, the largest continent, also has the most people. The Himalayas, the world's highest mountains, are in Asia. Africa is the second biggest continent. It has the largest desert and the longest river in the world. The Nile River flows over the high, dry land of the Sahara Desert.

Australia and Antarctica have water all around them. Australia is the smallest continent. The second biggest desert area is in Australia. Antarctica is the coldest continent. The South Pole is in Antarctica.

A. Write the letter for the correct answer.

1. Continents are the largest ____b____ masses.
   (a) water (b) land (c) ice

2. North America is connected to the continent of ____a____ .
   (a) South America (b) Africa (c) Europe

3. Antarctica and ____b____ have water all around them.
   (a) Asia (b) Australia (c) Africa

4. Europe, Asia, and ____c____ are joined together.
   (a) Antarctica (b) Australia (c) Africa

B. Draw lines to match the continent with its description.

1. North America   smallest continent
2. South America   coldest continent
3. Europe   longest coastline
4. Asia   hot weather and rain forests
5. Africa   jagged coastline and mild weather
6. Australia   largest continent and most people
7. Antarctica   largest desert and longest river

C. Fill in the missing words.

1. Many kinds of weather are found in ____North America____ .
   (Antarctica, North America)

2. The biggest lake in the world lies between ____Europe____ and Asia. (Australia, Europe)

3. Africa has the largest desert and the longest ____river____ in the world. (river, coastline)

4. Australia has the second biggest ____desert____ area. (desert, mountain range)

5. The South Pole is in ____Antarctica____ . (Australia, Antarctica)

Mountains

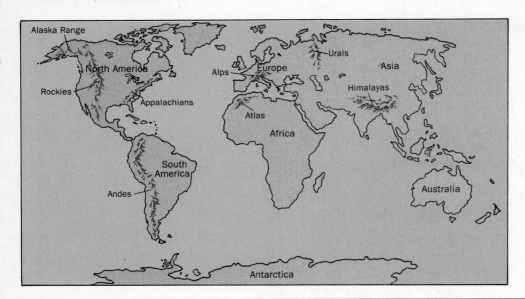

Mountains are landforms that are much higher than the land around them. To be a mountain, a landform must be 2,000 feet above the ground around it. The highest mountain on dry land is Mount Everest. It is in Asia. It is almost 30,000 feet high. The highest mountain in the United States is Mount McKinley, in Alaska. It is about 20,000 feet high.

Mountains have different shapes. Some are rounded and fairly low. These are usually older mountains that have been worn down over time. The Appalachian Mountains are an example of old mountains. Newer mountains are high and steep. They are rugged, like the Rocky Mountains.

All mountains have certain features. They have long **slopes,** or sides that curve down. The low places between mountains are called **valleys.** Some have deep valleys with steep sides called **canyons.** Many mountains have pointed tops called **peaks.** Others have long, narrow high places where two sloping surfaces meet called **ridges.**

Groups of mountains are called **ranges.** Some of the major mountain ranges include the Alaska Range, the Rockies, the Appalachians, the Andes, the Urals, the Alps, the Atlas, and the Himalayas. Many ranges are found along the edges of continents.

A. Answer True or False.

1. Mountains are landforms that are much higher than the land around them. ___True___

2. Mountains have different shapes. ___True___

3. Groups of mountains are called canyons. ___False___

4. Many mountain ranges are found along the edges of continents.
___True___

B. Fill in the missing words.

1. To be a mountain, a landform must be ___2,000___ feet above the ground around it. (200, 2,000)

2. The highest mountain on dry land is almost ___30,000___ feet high. (400, 30,000)

3. Some mountains are rounded and fairly ___low___. (low, steep)

4. Newer mountains are high and ___steep___. (flat, steep)

5. The Appalachian Mountains are an example of ___old___ mountains. (growing, old)

6. The low places between mountains are called ___valleys___. (ranges, valleys)

7. Many mountains have tops called ___peaks___. (peaks, canyons)

C. Draw lines to match each feature with its description.

1. valleys
2. slopes
3. peaks
4. canyons
5. ridges
6. mountain ranges

pointed tops of mountains

deep valleys with steep sides

low places between mountains

long sides that curve down

groups of mountains

long, narrow high places where two sloping surfaces meet

63

Plains and Plateaus

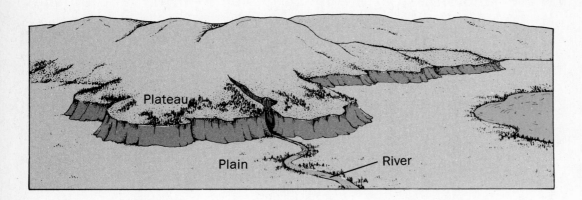

Plains and **plateaus** are large areas of flat land. A plain is usually lower than the land around it. Plateaus rise above the land around them. The edges of a plateau are often steep like cliffs.

The largest region in the United States is a plain. The Central Plains and the Great Plains reach from the Rocky Mountains to the Appalachian Mountains. This region has rich soil that is good for farming.

Plateaus such as those found in the western United States are covered with grass and are good for grazing sheep and cattle. Large deposits of coal can be found in the plateaus of the eastern United States. Fast-moving rivers have cut deep valleys or canyons in some plateaus. The Grand Canyon in Arizona is a canyon cut out of the Colorado Plateau.

Underline the correct words.

1. Plains and plateaus are large areas of (hilly, flat) land.

2. The largest region in the United States is a (plain, plateau).

3. A plain is usually (lower, higher) than the land around it.

4. The Central Plains and the Great Plains are good for (fishing, farming).

5. Large deposits of (water, coal) can be found in the plateaus of the eastern United States.

Part A

Fill in the missing words.

1. A natural flow of water that runs into a lake, ocean, or other body of water is a ____river____ . (river, lake)

2. Rainwater that flows across the ground is called _____run-off_____ . (ground water, run-off)

3. Ponds are often small and shallow enough that ____sunlight____ reaches the bottom. (water, sunlight)

4. To be a mountain, a landform must be ___2,000___ feet above the ground around it. (20, 2,000)

5. The low ground that holds a lake is called a ____basin____ . (plateau, basin)

Part B

Draw lines to match each feature with its description.

1. stalactites   flat lands rising above surrounding land

2. plateaus   wide, flat plain of the ocean floor

3. abyssal plain   rocks hanging from cave roofs

Part C

Write the letter for the correct answer.

1. Plankton are food for many ___b___ .
    (a) people   (b) animals   (c) plants

2. The largest land masses on Earth are called ___c___ .
    (a) caves   (b) plateaus   (c) continents

3. The rock found in most caves is ___b___ .
    (a) gravel   (b) limestone   (c) sand

4. Most of the water on Earth is found in the ___b___ .
    (a) rivers   (b) oceans   (c) lakes

Make a Model Island

You Need

- gallon plastic container
- water
- 6 toothpicks
- tape
- 1 stick modeling clay
- paper
- scissors

1. Review the model of the ocean floor on page 56. Which land feature is the island?

2. Cut one inch off the bottom of the container as shown.

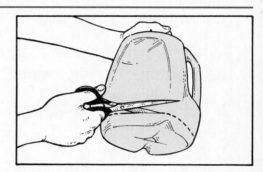

3. Mold the clay in the bottom of the container to form all of the ocean floor features. Why is the island about the same height as the shoreline?

4. Make labels for each land feature. Tape each label to a toothpick. Stick the labeled toothpicks into the clay to identify each feature.

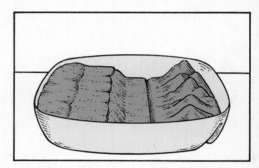

5. Carefully pour a small amount of water into the bottom of the container. Make sure the labels are above the water line. Which land features should be above the water? Which land features should be below the water?

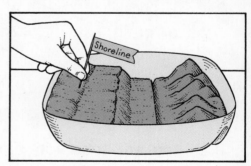

Save this model for use in Unit 4 activity.

Write the Answer

Describe what an island looks like above and below the water.

An island is a land formation surrounded by water.

It is actually a tall, underwater mountain peak

that sticks out of the water.

Fill in the circle in front of the word or phrase that best completes each sentence. The first one is done for you.

1. A river is a natural flow of water that runs
   - ● into a lake or an ocean.
   - ⓑ into the street.
   - ⓒ through many cities.

2. A hole that is dug to get ground water is a
   - ⓐ valley.
   - ⓑ canyon.
   - ⓒ well.

3. Most of the ocean floor is made up of the
   - ⓐ slope.
   - ⓑ trenches.
   - ⓒ abyssal plain.

4. Some of the largest and the smallest animals on Earth live in
   - ⓐ the oceans.
   - ⓑ caves.
   - ⓒ rivers.

5. The largest land masses on Earth are called
   - ⓐ caves.
   - ⓑ plateaus.
   - ⓒ continents.

6. To be a mountain, a landform must be
   - ⓐ 30,000 feet high.
   - ⓑ rounded and low.
   - ⓒ 2,000 feet above the ground around it.

Fill in the missing words.

7. The low ground that holds a lake is called a _____basin_____. (basin, well)

8. Most of the water on Earth is found in the _____oceans_____. (clouds, oceans)

9. Three fourths of Earth is covered with _____water_____. (plains, water)

Write the answer on the lines.

10. How are caves formed?

Caves are formed when ground water

wears away underground rock.

67

Climate Zones

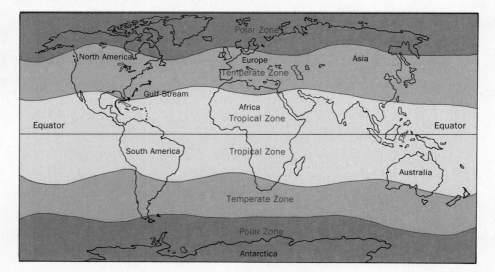

The weather in any one place can change from day to day. But the **climate** of a region does not change very much. Climate is the average weather of a region over a long period of time.

There are three main climate zones: **tropical, temperate,** and **polar.** In a tropical climate, it is warm all year round and rainfall is high. In a temperate climate, winters are cold and summers are warm. There is a <u>moderate</u> amount of rain. In a polar climate, it is cold much of the year and little rain or snow falls.

There are many factors that affect the climate of an area. Two of these factors are **latitude** and nearness to ocean **currents.**

Latitude is the distance north or south of the equator. The nearer a place is to the equator, the lower its latitude. Places near the equator have tropical climates. The places farthest from the equator have polar climates.

A current is a moving stream of water in an ocean. The <u>Gulf Stream</u> is a current of warm water. It moves north along the east coast of the United States. This current warms the nearby land areas. Cold ocean currents make land areas cooler.

A. Answer <u>True</u> or <u>False</u>.

1. The weather in any one place can change from day to day.

   _____True_____

2. The climate of a region changes a great deal. _____False_____

3. Places near the equator have polar climates. _____False_____

4. A current is a moving stream of water in an ocean. _____True_____

5. The Gulf Stream is a current of cold water. _____False_____

B. Fill in the missing words.

1. _____Climate_____ is the average weather of a region over a long period of time. (Climate, Latitude)

2. The places farthest from the equator have _____polar_____ climates. (tropical, polar)

3. _____Latitude_____ is the distance north or south of the equator. (Climate, Latitude)

4. In a _____temperate_____ climate, winters are cold and summers are warm. (temperate, tropical)

C. Name the three main climate zones.

1. _____tropical_____

2. _____temperate_____

3. _____polar_____

D. Answer the questions.

1. Look at the map on page 68. In what climate zone does most of the United States fall? _____Most of the United States is in the temperate climate zone._____

2. What are two of the factors that affect the climate of an area? _____Two of the factors are latitude and nearness to ocean currents._____

Deserts

A Desert

Deserts are places that get less than 10 inches of rain in a year. Most deserts are found in tropical climates. The largest desert is the Sahara in Africa.

Some deserts are covered by sand. The sand may pile up into tall **dunes.** Sand dunes are continually moved and shaped by the wind. Some deserts are covered by different kinds of rock and soil. Others have mountainous regions.

An **oasis** is an area in a desert that has water. The water may come from a spring or an underground stream. Palm trees and other plants grow near the water.

The plants and animals in a desert have adjusted to living in a climate with very little water. Desert plants such as cacti have small leaves or no leaves at all. This is because plants lose water through their leaves. Many desert plants have very long roots so that they can reach whatever water is available. The barrel cactus soaks up water in its stem after a rainfall. The stem becomes thinner as the cactus uses the water.

Most desert animals look for food at night and stay out of the hot sun during the day. Many small animals dig holes and stay underground in the daytime. Larger animals stay in shady places during the hottest part of the day.

The camel is a common animal in the deserts of Africa and Asia. A camel can go for a long time without water. If there is no food or water, a camel can live for a time on the fat stored in its hump.

A. Fill in the missing words.

1. Most deserts are found in ___tropical___ climates. (tropical, polar)

2. ___An oasis___ is an area in a desert that has water. (A mountain, An oasis)

3. Some deserts are covered by ___sand___. (water, sand)

4. Many desert plants have very long ___roots___. (stems, roots)

5. In a desert, sand may pile up into tall ___dunes___. (dunes, poles)

B. Answer True or False.

1. Deserts are places that get more than 100 inches of rain in a year. ___False___

2. All deserts are covered by sand. ___False___

3. The plants and animals in a desert have adjusted to living in a climate with very little water. ___True___

4. Desert animals look for food in the hot sun. ___False___

5. Cactus plants have large leaves. ___False___

C. Choose the word or words that best match each description.

cactus	deserts	oasis
camel	dunes	palm tree

1. places that get less than 10 inches of rain in a year ___deserts___

2. a kind of desert plant ___cactus___

3. an area in a desert that has water ___oasis___

4. a common animal in the deserts of Africa and Asia ___camel___

5. tall piles of desert sand ___dunes___

6. one kind of plant that grows near water in a desert ___palm tree___

71

Forests

Deciduous Forest

Rain Forest

Forests are places where many trees grow. They are found in both tropical and temperate climates. The climate and the kind of soil make forests different.

In tropical climates near the equator, it is warm and wet all year long. More than 100 inches of rain may fall during the year. **Tropical rain forests** grow in this climate. The largest tropical rain forest is along the Amazon River in South America.

More kinds of plants and animals are found in the tropical rain forest than anywhere else on Earth. Trees such as ebony, mahogany, and rosewood grow in the tropical rain forest. Vines grow up along the trunks of the trees. Orchids grow in the branches.

Some of the animals in a tropical rain forest are monkeys, lizards, and snakes. Many kinds of birds and insects make their homes there, too.

Most of the United States is in the temperate climate zone. On the West Coast, winters are mild and rainfall is heavy. Redwood and sequoia trees grow in forests in this area. These trees are **evergreens.** They stay green all year long. Redwoods and sequoias are the tallest trees on Earth. They are also some of the oldest.

In the northeastern part of the United States winters are cold and summers are warm. Evergreen and **deciduous forests** grow in this area. Deciduous trees lose all their leaves in the fall. Oak, beech, hickory, and maple are deciduous. Pine trees are evergreens.

A. **Answer True or False.**

1. Forests are places where many trees grow. _____True_____

2. Forests are found only in tropical climates. _____False_____

3. Most of the United States is in a tropical climate zone. _____False_____

4. The largest tropical rain forest is along the Amazon River in South America. _____True_____

5. Deciduous forests grow in the temperate climate zone. _____True_____

6. Evergreen trees stay green all year long. _____True_____

B. **Write the letter for the correct answer.**

1. The climate and the kind of ____b____ make forests different.
   (a) sunlight   (b) soil   (c) wind

2. More kinds of plants and animals are found in the ____a____ forest than anywhere else on Earth.
   (a) tropical rain   (b) deciduous   (c) evergreen

3. ____b____ trees lose all their leaves in the fall.
   (a) Tropical rain   (b) Deciduous   (c) Evergreen

4. On the West Coast of the United States, winters are mild and rainfall is ____c____ .
   (a) light   (b) moderate   (c) heavy

5. In tropical climates near the ____a____ , it is warm and wet all year.
   (a) equator   (b) poles   (c) United States

C. **Answer the questions.**

1. What is the climate like in the northeastern United States?

   Winters are cold and summers are warm.

2. What are two kinds of trees that grow in the tropical rain forest?

   Answers may vary, but should include: ebony,

   mahogany, and rosewood.

73

Grasslands

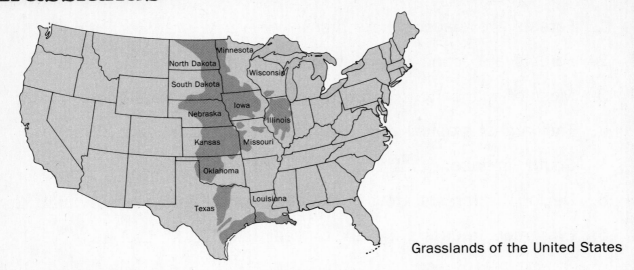

Grasslands of the United States

Grasslands are found in both temperate and tropical climates. Rainfall in the grasslands is between 10 and 20 inches a year. Because so little rain falls, few trees grow in the grasslands. But many kinds of grasses are found there. Some grasses grow wild. Others are planted by farmers. Many of these grasses are called **grains.** Wheat, corn, and oats are examples of grains.

Grasslands contain some of the best farmland in the world. They often have rich soils. The grasslands of the United States are also known as the Corn Belt and the Wheat Belt. Huge amounts of food are grown here.

Much of the midwestern United States is made up of grasslands. This area was once the home of herds of bison and antelope. But hunting and the spread of ranches cut down the number of these animals. Today, horses, cows, and sheep graze, or feed, on the grasslands.

Grasslands are also found in other parts of the world. In Australia, the grasslands are home for kangaroos. Some grasslands in Africa have enough water to support trees. Elephants, giraffes, zebras, and antelope feed on the plants and trees. Lions live on the grasslands, too. They feed on the antelope and other grazing animals.

A. Answer True or False.

1. Grasslands are found in both temperate and tropical climates.
   True

2. Rainfall in the grasslands is about 300 inches a year. _False_

3. Many trees grow in the grasslands. _False_

4. Many kinds of grasses are found in the grasslands. _True_

5. Much of the midwestern United States is made up of forests.
   False

6. Horses, cows, and sheep graze, or feed, on the grasslands.
   True

7. Grasslands often have poor soils. _False_

B. Use the words below to complete the sentences.

antelope	farmland	kangaroos	trees

1. Grasslands contain some of the best _farmland_ in the world.

2. In Australia, the grasslands are home for _kangaroos_.

3. Some grasslands in Africa have enough water to support _trees_.

4. The grasslands area of the midwestern United States was once the home of herds of bison and _antelope_.

C. Answer the questions.

1. Some grasses are planted by farmers. Many of those grasses are called grains. What are some of these grains? _Wheat, corn, and oats are examples of grains._

2. What are three animals that can be found in the grasslands of Africa?
   Answers may vary, but should include: elephants, giraffes, zebras, antelope, and lions.

75

Mountains

The Grand Tetons

You know that climates change as you go from the equator to the poles. Close to the equator there are tropical climates. Farther from the equator are the temperate and polar climate zones.

Climates can also change with **altitude,** or the height of an area above sea level. The higher the altitude of an area, the lower the average temperature of the area. So a very tall mountain could have more than one climate zone.

Imagine a tall mountain in a country near the equator. The climate at the bottom of the mountain would be tropical. A tropical rain forest might grow there. As you go up the mountain, the climate changes. Above the rain forest you would find a forest of pines and other evergreen trees.

Above the evergreen forest would be an area where no trees grow. This area is called an alpine meadow. Grasses and small flowering plants grow in the meadow.

Above the alpine meadow, the peak of the mountain is covered with snow for most or all of the year. This part of the mountain has a polar climate.

Deer and mountain sheep live high on the mountain during the summer. In winter, they go down closer to the base where they can find the plants they use for food.

Some birds, like the golden eagle, make nests high on the mountain. The golden eagle feeds on small animals and other birds.

A. Fill in the missing words.

1. Close to the equator there are _____tropical_____ climates.
   (tropical, polar)

2. Climates can change as you go from the equator to the
   _____poles_____. (poles, center of Earth)

3. _____Altitude_____ is the height of an area above sea level.
   (Altitude, Latitude)

4. The higher the altitude of an area, the _____lower_____ the average
   temperature of the area. (higher, lower)

5. An area high on a mountain where no trees grow is called an
   _____alpine meadow_____. (equator, alpine meadow)

B. The drawing below shows a mountain in a country near the equator. The climate changes as you go up the mountain. Tell what you would find at each level.

snow

alpine meadow

forest of evergreens

tropical rain forest

C. Answer the questions.

1. How do climates change as you go from the equator to the poles?

   Close to the equator there are tropical climates. Farther

   from the equator are the temperate and polar climate zones.

2. How does climate change with altitude? _____The higher

   the altitude of an area, the lower the average

   temperature of the area.

The Arctic

Arctic Tundra

The area around the North Pole is known as the **Arctic.** This area includes the northern parts of Europe, Asia, and North America. The Arctic has a polar climate. In winter, the average temperature is about 30 degrees below 0°F. The average summer temperature is about 50°F.

Much of the Arctic is covered by **tundra.** Tundra is a huge area of plains with no trees. In the summer, mosses, lichens, grasses, and many flowering plants grow in the tundra. But the soil below the surface of the ground is always frozen.

Reindeer and caribou are the most common animals in the Arctic. They feed on grasses, shrubs, and lichens. In winter, they go south of the tundra to find food. Bears, wolves, foxes, and rabbits also live in the Arctic. Birds are common, too. Ducks, geese, owls, and puffins are a few of the birds that make their home there.

Answer True or False.

1. The Arctic has a tropical climate. _____False_____

2. No plants grow in the Arctic. _____False_____

3. Reindeer and caribou are the most common animals in the Arctic.

   _____True_____

Part A

Use the words below to complete the sentences.

altitude	equator	temperate
climate	Forests	tropical
current	grasslands	tundra
deserts	oasis	

1. The average weather of a region over a long period of time is called ___climate___ .

2. Latitude is the distance north or south of the ___equator___ .

3. Places near the equator have ___tropical___ climates.

4. Places that get less than 10 inches of rain in a year are ___deserts___ .

5. An area in a desert with water is called an ___oasis___ .

6. ___Forests___ are places where many trees grow.

7. Deciduous forests grow in the ___temperate___ climate zone.

8. Much of the midwestern United States is made up of ___grasslands___ .

9. The higher the ___altitude___ of an area, the lower the average temperature of an area.

10. Some of the land in the Arctic is ___tundra___ .

Part B

Draw lines to match each climate zone with its description.

1. temperate   warm all year round

2. tropical   winters are cold and summers are warm

3. polar   cold much of the year

Make a Matching Map

You Need

- a partner
- pencil
- paper
- clay and container bottom from Unit 3 activity

1. Look at the map on this page. How does the map show mountains? How does the map show rivers and lakes? Are there other land forms too? Work with your partner to locate as many land forms as you can find on the map.

2. Separate your clay in two. Use a little to put a thin layer in the bottom of the container. Use some more to make a mountain like the one on the map.

3. Use your finger to scoop out the lake. Use a pencil to make the riverbed.

4. Work with your partner to use some clay to build up the plateau. What is the land of your place like? Check that your model has all the land forms in the map.

5. Trace the map on this page. Hold it over your model. Does the model look like the map? Make the model match the map.

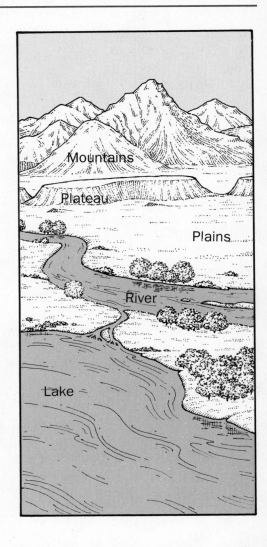

Write the Answer

Describe how a map shows where mountains, lakes, and rivers are.

Answers will vary. All answers should indicate that

the student has used the map legend to

identify the different features.

Fill in the circle in front of the word or phrase that best completes each sentence. The first one is done for you.

1. The average weather of a region over a long period of time is
   ⓐ tundra.
   ● climate.
   ⓒ current.

2. Altitude is the height of an area above
   ⓐ the tree line.
   ⓑ the equator.
   ⓒ sea level.

3. The Arctic includes the
   ⓐ North Pole.
   ⓑ United States.
   ⓒ equator.

4. The top of a very high mountain may be covered by
   ⓐ sand.
   ⓑ snow.
   ⓒ tropical rain forests.

5. The distance north or south of the equator is
   ⓐ altitude.
   ⓑ height.
   ⓒ latitude.

6. Grasslands are found in
   ⓐ deserts.
   ⓑ polar climates.
   ⓒ both temperate and tropical climates.

Fill in the missing words.

7. Some grasslands are used to grow _____oats_____. (fruit, oats)

8. Places farthest from the equator have _____polar_____ climates. (polar, tropical)

9. Rain forests grow in _____tropical_____ climates. (tropical, temperate)

Write the answer on the lines.

10. Describe the climate of a tropical rain forest.

   The climate of a tropical rain forest
   _____
   is warm and wet all year.
   _____

UNIT 5
Shaping the Surface

Bryce Canyon

Weathering and Erosion

The surface of Earth is always being shaped and worn away by water, wind, and ice. This is a natural process called **erosion.** Erosion is the wearing down and moving of rocks and soil from one place to another.

Erosion begins with **weathering.** Weathering is the breaking up of rocks and soil on Earth's surface. The surface is made of many different kinds of rock. As the rock becomes exposed on the surface, it is slowly broken into smaller pieces by wind, rain, and ice.

Erosion takes place when the small pieces of rock and soil are carried away by the wind or moving water. Erosion is usually so slow that you cannot tell it is happening. At other times, landslides, mudslides, or storms cause erosion to happen very quickly.

The activities of people can add to weathering and erosion. When people clear land for farms, roads, and cities, we expose the surface. This can cause soil erosion. Trees and other plants protect the soil from eroding. Plant roots hold the soil in place. They also help to absorb rain water that can wash away soil.

Weathering and erosion can be helpful or harmful. Weathering makes new soil by breaking up rock. Many beautiful landforms, like Bryce Canyon in Utah, were formed by weathering and erosion. But weathering and erosion can carry away rich farm soil. Eroded soil can clog ditches and streams and cause flooding.

A. Answer True or False.

1. The surface of Earth is always being shaped. ___True___

2. Erosion is the wearing down and moving of rocks and soil from one place to another. ___True___

3. Erosion always happens very quickly. ___False___

4. The activities of people can add to weathering and erosion. ___True___

5. Weathering and erosion can carry away rich farm soil. ___True___

B. Write the letter for the correct answer.

1. Erosion begins with ___b___.
   (a) people   (b) weathering   (c) mudslides

2. Weathering is the ___a___ of rocks and soil on Earth's surface.
   (a) breaking up   (b) forming   (c) moving

3. Trees and other plants protect the soil from ___c___.
   (a) insects   (b) ice   (c) erosion

4. As rock becomes exposed on the surface, it is slowly broken into ___b___ by wind, rain, and ice.
   (a) huge pieces   (b) smaller pieces   (c) dust

5. At times, landslides, mudslides, or ___c___ cause erosion to happen very quickly.
   (a) rocks   (b) ice   (c) storms

C. Draw lines to match each word with its description.

1. Earth's surface   the moving of rocks and soil

2. erosion   made of many different kinds of rock

3. weathering   formed by weathering and erosion

4. Bryce Canyon   breaking up of rocks and soil

5. eroded soil ———————— clogs ditches and streams

83

Water Changes the Surface

Soil Deposits in a River Delta

Canyon Formed by the Rio Grande

Moving water is the major cause of weathering and erosion on Earth. Rain, rivers, and ocean waves are examples of moving water.

Rain causes weathering of rocks and soil. Some rocks, like <u>limestone</u>, may break down or <u>dissolve</u> in water. Little by little, rain dissolves the rock and washes it away.

As rain splashes against the ground, it loosens tiny bits of soil. When rain begins to flow across the ground, it is called <u>run-off</u>. Run-off flows from small paths of water down into larger paths of water. The run-off carries the loosened soil into rivers.

Water that flows in rivers wears away the land and moves rocks and soil downhill. Flowing water may wear away the <u>riverbed</u> and make a **V-shaped canyon.** Flowing water may also cut away the banks of a river. Then the river begins to flow in curves. Often the water in the middle of rivers flows slowly. Slow-flowing water can deposit the rocks and soil it has picked up. These deposits can make a landform like the <u>Mississippi</u> <u>River</u> <u>Delta</u>.

As a river flows into the ocean, it drops some of the material it is carrying. The water in oceans is always moving. This makes **waves** that pound against the shore. Waves wear away the land, making steep cliffs. Ocean waves can deposit the sand and soil they are carrying. These deposits can make **sandbars** near the shore. They also make sandy beaches.

A. Answer True or False.

1. Moving water never causes erosion. **False**

2. Rain causes weathering of rocks and soil. **True**

3. When rain soaks into the ground, it is called run-off. **False**

4. Run-off carries loosened soil into rivers. **True**

5. Water that flows in rivers never wears away the land. **False**

6. Flowing water may cut away the banks of a river. **True**

7. The water in oceans is always moving. **True**

B. Fill in the missing words.

1. Some rocks, like limestone, may break down or **dissolve** in water. (dissolve, get stronger)

2. Flowing water may **break down** the riverbed and make a V-shaped canyon. (build up, break down)

3. Ocean waves can deposit sand and soil to make **sandbars** and sandy beaches. (canyons, sandbars)

4. Often the water in the middle of rivers flows **slowly**. (slowly, very fast)

5. Water that flows in rivers moves rocks and soil **downhill**. (uphill, downhill)

C. Draw lines to match each word with its description.

1. limestone — dissolves in water
2. run-off — water that flows across ground
3. ocean waves — deposits made by waves
4. flowing water — cuts away the banks of a river
5. sandbars — make steep cliffs
6. Mississippi River Delta — landform made by river deposits

Ice Changes the Surface

The ice from these glaciers moves rocks and soil.

Water in the form of frost, hail, snow, and ice changes the surface of Earth in many ways.

Frost is frozen dew, or small drops of water. When water freezes, it takes up more space. If frost forms in the cracks of a rock, it pushes on both sides of the crack. Then the crack gets bigger.

Hail and snow fall to Earth when water freezes in the air. Hailstones are hard, round pieces of ice. Hailstones pound at the ground and loosen soil. Snow is frozen water that falls to Earth in soft flakes. Snow may cover large areas of Earth. When hail and snow melt, water may flow across the ground as run-off. Run-off moves soil.

In polar regions and at the top of some mountains, water forms large, slow-moving sheets of ice called **glaciers.** These sheets of ice grind up and move rock and soil. Glaciers are so big and heavy that they press down very hard on the land. Rocks and soil get stuck on the bottom of glaciers.

As a glacier moves, the rocks and soil in the lower layer act like sandpaper against Earth. They scratch and grind down the soil. In this way, glaciers cut out **U-shaped valleys.**

When glaciers melt, some of the rocks and soil they were carrying are left behind. Some of this material forms ridges called **moraines.** Sometimes, the rocks and soils moved by a glacier form a whole island, such as Long Island in New York.

A. **Answer True or False.**

1. Water in the form of frost, hail, snow, and ice changes the surface of Earth. ___True___

2. Frost makes the cracks in a rock get smaller. ___False___

3. Hailstones are large sheets of ice and snow. ___False___

4. Glaciers grind up and move rock and soil. ___True___

5. As a glacier moves, the rocks and soil in the lower layers act like sandpaper against Earth. ___True___

B. **Write the letter for the correct answer.**

1. When water freezes, it takes up ___b___ space.
   (a) less   (b) more   (c) lighter

2. Frost makes the cracks of a rock get ___b___ .
   (a) smaller   (b) bigger   (c) smoother

3. Hailstones ___c___ at the ground and loosen soil.
   (a) grind   (b) flow   (c) pound

4. Glaciers are so ___a___ that they press down very hard on the land.
   (a) big and heavy   (b) small and light   (c) fast and smooth

5. Glaciers cut out ___b___ valleys.
   (a) V-shaped   (b) U-shaped   (c) L-shaped

6. Some of the material left behind by glaciers forms ___b___ .
   (a) valleys   (b) moraines   (c) layers

C. **Draw lines to match the word with its description.**

1. frost   soft flakes

2. glacier   frozen dew

3. snow   hard, round pieces of ice

4. hailstones   large sheet of slow-moving ice

Wind Changes the Surface

Cap Rock in the Badlands

Hurricane Damage

Wind is air moving across Earth's surface. Winds can be gentle. They can also blow so hard that they can damage buildings and push over large trees. In strong winds, tiny particles of rock and soil are picked up and carried away. Particles carried by the wind wear away rock and change Earth's surface.

As wind blows, the rock and soil particles hit against exposed rock. This action is like sandpaper wearing away wood. Rocks can be carved into strange shapes. Cap Rock in the Badlands of South Dakota was formed this way.

Wind can work with rain to weather and erode Earth's surface. Wind pushes rain so it hits Earth harder. Together wind and rain can loosen and move more soil than they can separately.

Strong winds and heavy rain may form a hurricane over the ocean. When a hurricane moves onto land, it can do a great deal of damage. Pounding ocean waves can destroy beaches. The strong winds can also push over large trees that hold the soil. Then heavy rains can wash the soil away.

Earth's surface can also be shaped by wind and snow. The strong winds of a blizzard can push over trees and blow snow into deep drifts that are very heavy. Rocks and soil get stuck in the snow. When the snowdrifts melt, the flowing water can move the rocks and soil, causing erosion and flooding.

A. Answer True or False.

1. Wind is air moving across Earth's surface. _____True_____

2. Wind pushes rain so it does not hit Earth. _____False_____

3. Strong winds and deep snow may form a hurricane. _____False_____

4. In strong winds, tiny particles of rock and soil are picked up and carried away. _____True_____

5. The strong winds of a blizzard can blow snow into deep drifts that are very heavy. _____True_____

B. Fill in the missing words.

1. Winds can blow so hard and fast that they can damage buildings and push over large _____trees_____. (mountains, trees)

2. As wind blows, the rock and soil particles _____hit_____ against exposed rock. (float, hit)

3. In a hurricane, pounding ocean waves can _____destroy_____ beaches. (destroy, save)

4. Strong winds and heavy rain may form a _____hurricane_____ over the ocean. (hurricane, blizzard)

5. Together wind and rain can loosen and move _____more_____ soil than they can separately. (less, more)

6. When snowdrifts melt, the flowing water can move the rocks and soil, causing _____erosion_____ and flooding. (blizzards, erosion)

C. Use each word to write a sentence about wind.

1. rock _____Sentences will vary._____

_____

2. blizzard_____

_____

89

Soil

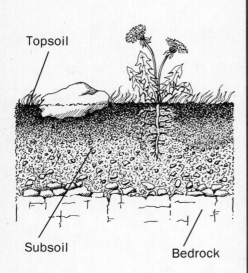

Topsoil

Subsoil

Bedrock

Soil is made of rocks that slowly break down into tiny bits. These bits of broken rock mix with air, water, and **humus** to form soil. Humus is the remains of dead plants and animals. Soil forms slowly. One inch of soil takes hundreds or thousands of years to form.

There are many different kinds of soils. That is because different areas have rocks made of different minerals. Each kind of rock makes a certain type of soil. Feldspar makes **sand.** Sand has the biggest grains, or bits. It is very coarse. When quartz is broken down, it makes **silt.** Silt grains are tiny but they can be seen. Micas and other minerals break down to make **clay.** Clay has such tiny grains that one grain cannot be seen without a microscope.

Soil often has two layers. The layers are the **topsoil** and the **subsoil.** The topsoil is usually the richest part of the soil. It has rock particles, humus, and minerals. In this layer, dissolved minerals and nutrients from the humus make it possible for plants to grow. The subsoil is made of larger rocks that have been broken down. Solid rock called **bedrock** is under the layers of soil.

Erosion from wind and water carry away tons of soil each year. One way that farmers protect soil is by planting rows of crops around a slope, instead of up and down. The water soaks into the ground instead of washing away soil. People also put grass and plants in places where the soil is bare. Plants help to hold the soil. Wind and water will not be able to move the soil so easily.

A. Answer True or False.

1. Soil is made of rocks that slowly break down into tiny bits.
   True

2. Humus is living plants and animals. **False**

3. One inch of soil takes hundreds or thousands of years to form.
   True

4. There is only one kind of soil. **False**

5. Each kind of rock makes a certain type of soil. **True**

6. Topsoil is usually the richest part of the soil. **True**

B. Fill in the missing words.

1. Bits of broken rock mix with air, water, and **humus** to form soil. (dust, humus)

2. Soil forms **slowly**. (slowly, very fast)

3. Solid rock called **bedrock** is under the layers of soil. (bedrock, humus)

4. Micas and other minerals break down to make **clay**. (clay, quartz)

5. Soil often has **two** layers. (two, eight)

6. Erosion from wind and **water** carry away tons of soil each year. (dust, water)

C. Use each word to write a sentence about soil.

Sentences will vary.

1. sand _____

   _____

2. silt _____

   _____

3. clay _____

   _____

People Change the Surface

Strip Mining for Coal

People have changed Earth's surface in many ways. Farming, logging, fires, and mining are some of the ways that people change Earth's surface.

Three hundred years ago, one half of the United States was covered by forests. Forests are important in preventing erosion. The roots of trees help hold soil in place. Soil also holds a lot of water.

When settlers came, they cleared land for farms and towns by cutting down many trees. Today, only one third of the United States is covered by forests.

Logging companies often clear large areas of trees to be used as lumber. Without trees to hold water in the soil, rapid run-off causes flooding. Forests are also destroyed by fire. About three million acres of forests are burned each year in the United States. Most of these fires are started by people.

Coal is a mineral that is sometimes found just below the surface. Mining companies often remove the topsoil to get the coal. This kind of mining causes the land to erode quickly.

Underline the correct words.

1. People have changed Earth's surface in (many, few) ways.

2. Three hundred years ago, (all, one half) of the United States was covered by forests.

3. Rapid run-off causes (flooding, fires).

4. Most forest fires are started by (animals, people).

UNIT 5 Review

Part A

Use the words below to complete the sentences.

erosion	hurricanes	surface
glacier	rivers	wear away
humus	run-off	weathering

1. The _____surface_____ of Earth is always being worn away by water, wind, and ice.

2. The wearing down and moving of rocks and soil from one place to another is called _____erosion_____ .

3. The breaking up of rocks and soil on Earth's surface is caused by _____weathering_____ .

4. Rocks and soil are moved downhill by _____rivers_____ .

5. Large, slow-moving sheets of ice make up a _____glacier_____ .

6. Rainwater that flows across the ground is called _____run-off_____ .

7. Strong winds and heavy rain that form over the ocean are called _____hurricanes_____ .

8. Tiny bits of broken rock mix with air, water, and _____humus_____ to form soil.

Part B

Draw lines to match each word with its description.

1. frost ——————— frozen dew

2. hailstone — large, slow-moving sheet of ice

3. glacier — hard, round piece of ice

4. topsoil — deposit of sand near shore

5. sandbar — often the richest layer of soil

93

Show Weathering Effects

You Need

- a partner
- 1 cup damp sand
- pie pan
- clean, small milk carton
- water
- bucket

1. Review the differences in mountains (page 62) and plateaus (page 64).

2. Work with a partner. Place the sand in the pan. Shape it into a mountain.

3. Fill the carton with water. Gently pour the water over the top of the mountain. How did the water change the shape of the mountain? How does rain change the shape of a mountain?

4. Pour the water out of the pan and into a bucket. Make a plateau out of the sand.

5. Fill the carton with water again. Gently pour the water over the plateau. How did the water change the shape of the plateau? How does rain change the shape of a plateau? How are these changes in mountains and plateaus the same and different?

Write the Answer

Describe how rain changes a mountain.

Rain splashes on the mountain and loosens the soil and the rocks.

The moving water carries these pieces away and over time flattens the

peak of the mountain. Water also runs off mountains to cut rivers and canyons.

Fill in the circle in front of the word or phrase that best completes each sentence. The first one is done for you.

1. Moving water wears away land and moves rocks and soil
   - ● downhill.
   - ⓑ uphill.
   - ⓒ into the mountains.

2. Frost helps break up rocks by pushing against the
   - ⓐ shore.
   - ⓑ banks.
   - ● cracks.

3. Soil is made of broken rocks, air, water, and
   - ⓐ sandbars.
   - ⓑ sheets of ice.
   - ● humus.

4. Rocks and soil blown by the wind can
   - ● wear away rock.
   - ⓑ form snowdrifts.
   - ⓒ form the oceans.

5. Each kind of rock makes a certain type of
   - ● soil.
   - ⓑ nutrient.
   - ⓒ humus.

6. People help cause weathering and erosion by
   - ⓐ planting trees.
   - ● clearing the land.
   - ⓒ growing grass.

Fill in the missing words.

7. Glaciers are large, slow-moving sheets of ____ice____.
   (sand, ice)

8. Erosion is the ____moving____ of rocks and soil.
   (moving, breaking up)

9. Weathering is the ____breaking up____ of rocks and soil.
   (moving, breaking up)

Write the answer on the lines.

10. What causes most weathering and erosion on Earth?

   Most weathering and erosion on Earth

   is caused by moving water.

UNIT 6
The Changing Earth

The Structure of Earth

Earth is shaped like a ball and is made up of four layers: the **inner core,** the **outer core,** the **mantle,** and the **crust.** It is about 8,000 miles in diameter.

At the very center of Earth is the inner core. It is about 800 miles thick and is made of the metals iron and nickel. The inner core is very hot, about 9,000° F. Although it is hot enough to melt, the inner core is solid because it is under so much pressure from the layers above. The molecules of iron and nickel cannot spread out enough to become liquid.

The outer core is about 1,400 miles thick. It is also made of iron and nickel. The temperature of the outer core ranges from 4,000° F to 9,000° F. Because the outer core is not under as much pressure as the inner core, the iron and nickel have melted. The outer core is made of hot liquid metal.

The mantle is the thickest layer of Earth. It is about 1,800 miles thick and lies just below Earth's crust. The mantle seems to be made up of several layers. The top layer of the mantle is hot solid rock. The bottom layer is liquid rock. The rocks in the mantle are made mostly of silicon, aluminum, iron, and magnesium.

The crust is the outer layer of Earth. It is about 20 miles thick in most places. But it can be as little as 5 miles thick. The crust makes up the continents and ocean floors. It is made up of many kinds of rock, such as granite, sandstone, and marble.

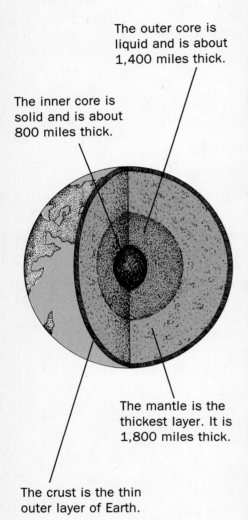

The outer core is liquid and is about 1,400 miles thick.

The inner core is solid and is about 800 miles thick.

The mantle is the thickest layer. It is 1,800 miles thick.

The crust is the thin outer layer of Earth.

A. Answer True or False.

1. Earth is made up of four layers. __True__

2. The inner core is made of iron and nickel. __True__

3. The outer core is made of solid rock. __False__

4. The mantle is the thickest layer of Earth. __True__

5. The crust is at the center of Earth. __False__

B. Draw lines to match each layer of Earth with its description.

1. inner core   is hot liquid metal

2. outer core   at the very center of Earth

3. mantle   makes up continents and ocean floors

4. crust   the thickest layer

5. Earth ———— 8,000 miles in diameter

C. Write the letter for the correct answer.

1. The ___c___ is made up of many kinds of rock, such as granite, sandstone, and marble.
   (a) inner core   (b) outer core   (c) crust

2. The ___a___ is very hot, about 9,000° F.
   (a) inner core   (b) mantle   (c) crust

3. The ___b___ is about 1,800 miles thick.
   (a) inner core   (b) mantle   (c) crust

4. The ___a___ is about 20 miles thick in most places, but it can be as little as 5 miles thick.
   (a) crust   (b) mantle   (c) inner core

5. The inner core is ___b___ rock because it is under so much pressure from above.
   (a) melted   (b) solid   (c) gaseous

6. The ___a___ seems to be made up of several layers.
   (a) mantle   (b) crust   (c) inner core

Plates of Earth

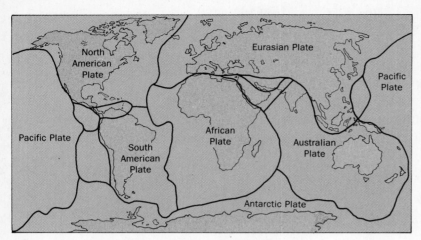

Earth's Seven Major Plates

The surface of Earth is made up of large sections called **plates.** These plates are like pieces of a puzzle that fit together. There are about 20 plates that make up Earth's crust. The seven major plates are the Pacific, the North American, the South American, the African, the Eurasian, the Australian, and the Antarctic.

The plates of the crust rest on the hot liquid rock of the mantle below them. The plates move slowly all the time. They are carried along as the melted rock in the mantle moves. Most plates move from $\frac{1}{4}$ inch to 4 inches a year.

Plates can move in several ways. They can move toward each other, away from each other, or past each other. Many of Earth's features are formed by the motion of plates.

When two plates under the ocean floor collide, one plate is pushed down below the edge of the other. A deep ocean **trench** is formed. When plates under the continents collide, the edges of the plates crumple and are pushed up. Mountains such as the Himalayas and Andes were formed in this way.

In other places, plates grind past each other. The line where two plates pass each other is called a **fault.** Strain is produced on the rocks on both sides of the fault. Many earthquakes occur along faults.

A. Answer True or False.

1. The surface of Earth is made up of large sections called plates.

   _____True_____

2. None of the plates on Earth move. _____False_____

3. The line between plates is called a trench. _____False_____

4. Plates move toward each other, away from each other, or past

   each other. _____True_____

5. There are seven major plates that make up Earth's crust. _____True_____

B. Fill in the missing words.

1. Many of Earth's features are formed by the motion of _____plates_____.
   (oceans, plates)

2. The plates _____rest_____ on the hot liquid rock of the mantle below
   them. (rest, burn)

3. Most plates move from _____$\frac{1}{4}$ inch to 4 inches_____ a year.
   ($\frac{1}{4}$ inch to 4 inches, $\frac{1}{4}$ mile to 4 miles)

4. The line where two plates pass each other is called a _____fault_____.
   (fault, ridge)

5. The Eurasian, the Africa, and the South American are three of the

   major _____plates_____ on Earth. (plates, crusts)

C. Use each word to write a sentence about the plates of Earth.

1. plate _____ Sentences will vary. _____

   _____

2. ocean trench _____

   _____

3. mountains _____

   _____

99

How Mountains Are Formed

Dome Mountains

Fold Mountains

Fault-Block Mountains

It takes millions of years for mountains to form. Even though they rise high above the ground, mountains begin forming below the surface of Earth.

Mountains can be formed when plates that make up Earth's crust move. Some mountains are formed when two plates push against each other. The pressure causes one plate to bulge out. Often the bulge is round and is lifted into the shape of a dome. **Dome-shaped mountains** are usually low mountains.

Fold mountains are formed when plates collide head-on. The crust folds and forms peaks. Sometimes when plates collide, the rock breaks into blocks. It is pushed up along fault lines. These broken blocks form **fault-block mountains.** Fault-block mountains, like the Sierra Nevada in North America, are often steeper than other kinds of mountains.

Some mountains begin when one plate is forced down below another plate. The rock in the lower plate melts then rises to the surface. This hot melted rock, or **lava,** spews out of the ground. As the melted rock cools, it forms a hard layer. More layers are formed as the lava piles up and cools. In this way, a **volcanic mountain** is formed.

Most mountains are formed in a group, or range. Mountains in a range have about the same shape and are usually made of the same kinds of rock. The longest range in the world is the Andes. It is found along the edge of the South American Plate.

A. Answer True or False.

1. It takes millions of years for mountains to form. ___True___

2. Mountains can be formed when plates that make up Earth's crust move. ___True___

3. Fold mountains are formed when plates collide head-on. ___True___

4. When rock breaks into blocks, volcanoes are formed. ___False___

5. A volcanic mountain can be formed when one plate is forced down. ___True___

6. Mountains begin forming below the surface of Earth. ___True___

7. As melted rock cools, it forms a liquid. ___False___

B. Draw lines to match the kind of mountain with its description.

1. dome-shaped   formed when rocks break into blocks

2. fold   made when a plate bulges

3. fault-block   formed when lava cools and piles up

4. volcanic   crust folds to form peaks

C. Write the letter for the correct answer.

1. Most mountains are formed in groups called ___a___.
   (a) ranges   (b) faults   (c) plates

2. Dome-shaped mountains are usually ___a___ mountains.
   (a) low   (b) high   (c) jagged

3. Mountains in a range have about the same ___b___.
   (a) size   (b) shape   (c) piles

4. The longest mountain range in the world is the ___b___.
   (a) Sierra Nevada   (b) Andes   (c) Rockies

5. Fault-block mountains are formed when broken rock is ___a___ along a fault line.
   (a) pushed up   (b) forced down   (c) melted

Earthquakes

An earthquake destroyed this building and caused the ground to split open.

An **earthquake** is a shaking of the surface of Earth when the plates of Earth move. About one million earthquakes happen every year. But most happen under the ocean or are too small to notice. In a strong earthquake, however, the ground can shake violently and even split open.

Most earthquakes are caused by the sudden movement of plates along a fault. A fault is the place where two plates meet. When the plates move, the crust near the fault is squeezed and stretched. Rocks in the crust break and move apart at a point called the **focus.** The focus is usually less than 25 miles below the surface.

The place on the surface of Earth that is directly above the focus is called the **epicenter.** Places near the epicenter receive the most damage.

If the focus of an earthquake is under the ocean floor, a **tsunami,** or giant wave, is produced. This huge wall of water moves across the ocean at high speed. It can cause great damage when it reaches land.

Many earthquakes take place in an area known as the Ring of Fire. Volcanoes are also common in this area. This chain of earthquake zones and volcanoes circles the Pacific Plate.

A. Answer True or False.

1. An earthquake is a shaking of the surface of Earth. _____True_____

2. A fault is the place where two plates meet. _____True_____

3. Most earthquakes are caused by the sudden movement of plates along a fault. _____True_____

4. The focus is where rocks stop breaking and moving in an earthquake. _____False_____

5. A tsunami is a tiny ocean wave. _____False_____

B. Fill in the missing words.

1. About one _____million_____ earthquakes happen every year. (hundred, million)

2. Most earthquakes are too _____small_____ to notice. (small, big)

3. A fault is the place where two _____plates_____ meet. (plates, epicenters)

4. In a _____strong_____ earthquake, the ground can shake violently and even split open. (small, strong)

5. When the plates move, the crust near the _____fault_____ is squeezed and stretched. (fault, tsunami)

6. The _____focus_____ is usually less than 25 miles below the surface. (epicenter, focus)

7. The Ring of Fire is a chain of _____earthquake_____ zones and volcanoes that circles the Pacific Plate. (mountain, earthquake)

C. Use each word to write a sentence about earthquakes.

1. epicenter _____ Sentences will vary. _____

_____

2. focus _____

_____

Volcanoes

An Erupting Volcano

When the plates of Earth collide, sometimes a **volcano** is formed. A volcano occurs when material from the mantle **erupts,** or blasts, onto the surface. Liquid rock, hot gases, pieces of rock, or a mixture of these may erupt from the volcano.

Not all volcanoes erupt in the same way. Some erupt quietly, while others erupt violently. In a quiet eruption, lava flows slowly out of the volcano. But sometimes, the **vent,** or opening, in the volcano is plugged. Then pressure builds up inside the volcano, and a violent eruption takes place. In 1980, Mount St. Helens in the state of Washington erupted violently. The blast blew 1,200 feet off the top of the mountain.

Different types of volcanic mountains are formed depending on the kind of eruption and the materials that pour out of the volcano.

Shield volcanoes are formed when lava spreads out around the vent in a quiet eruption. A low, rounded mountain, like Mauna Loa in Hawaii, is formed.

Cinder cones are made when a volcano erupts violently and throws cinders out the top. **Cinders** are volcanic rocks about the size of golf balls. The cinders fall around the vent and form a cone-shaped mountain.

A **composite volcano** is formed when cinders make up some layers and lava flows make up other layers. Composite volcanoes are towering cone-shaped mountains, like Mount Fuji in Japan.

A. **Answer True or False.**

1. When the plates of Earth collide, sometimes a volcano is formed.

   True

2. A volcano melts when rocks and other materials pour out.

   False

3. Some volcanoes erupt quietly. _True_

4. Different types of volcanoes form depending on the kind of eruption and the materials that pour out. _True_

B. **Use the words below to complete the sentences.**

cinder	lava	shield volcano
cinder cone	mountains	vent
composite volcano	pressure	volcano

1. A _volcano_ occurs when material from the mantle erupts onto the surface.

2. In a quiet eruption, _lava_ flows slowly out of the volcano.

3. A volcanic rock about the size of a golf ball is a _cinder_ .

4. A _shield volcano_ is formed when lava spreads out around the vent.

5. A _cinder cone_ is made when a volcano erupts violently and throws cinders out the top.

6. A _composite volcano_ is formed when cinders make up some layers and lava flows make up other layers.

7. A _vent_ is an opening in a volcano.

8. When a vent is plugged, _pressure_ builds up inside the volcano and a violent eruption takes place.

Energy Inside Earth

Old Faithful

The heat inside Earth is called **geothermal energy.** Hot springs and **geysers** are sources of geothermal energy. In a hot spring, hot water and gases bubble up and escape from the ground.

In a geyser, the hot water and steam shoot out of the ground like a volcanic eruption. Geysers are found in areas where there are cracks in Earth's crust. Water drains into these cracks until it reaches the rocks below, which are very hot. This water changes to steam. The steam lifts water above it. When enough steam is formed, water erupts out of the opening in Earth's surface in a rush. Huge columns of water shoot up from some geysers.

Old Faithful, a geyser in Yellowstone National Park in Wyoming, erupts about every 73 minutes. It can send water and steam 150 feet into the air.

Geothermal energy can be used to heat homes and to produce electricity. Today, geothermal energy heats most of the homes in Iceland. Scientists are looking for more ways to use this energy.

Underline the correct words.

1. In a hot spring, hot water and (rocks, <u>gases</u>) escape from the ground.

2. In a geyser, the hot water and steam (<u>shoot</u>, flow) out of the ground.

3. Geothermal energy can be used to heat homes and to produce (noise, <u>electricity</u>).

4. Hot springs are sources of (solar, <u>geothermal</u>) energy.

Part A

Use the words below to complete the sentences.

features	mantle	plates
fault	Mountains	range
hot spring	nickel	volcano

1. The inner core, the outer core, the _____**mantle**_____, and the crust are the four layers of Earth.

2. The inner core and the outer core are made of iron and _____**nickel**_____.

3. There are about 20 _____**plates**_____ that make up the crust of Earth.

4. Many of Earth's _____**features**_____ are formed by the motion of plates.

5. _____**Mountains**_____ can be formed when plates move.

6. Most earthquakes are caused by the sudden movement of plates along a _____**fault**_____.

7. A _____**volcano**_____ is an opening in Earth's crust where material from the mantle erupts onto the surface.

8. Most mountains are formed in a group, or _____**range**_____.

Part B

Draw lines to match each term with its description.

1. earthquake — shaking of the surface of Earth

2. focus — eruption of hot water and steam

3. geyser — made from lava flows and cinders

4. geothermal energy — place underground where rocks break apart

5. composite volcano — heat energy from inside Earth

Show How Mountains Form

You Need
- 2 felt squares each of red, yellow, and blue
- unlined paper
- red, yellow, and blue colored pencils

1. Review the different types of mountains on page 100.

2. Make two stacks of felt. Start with the red felt on the bottom, the yellow felt in the middle, and the blue felt on the top. These stacks will be the plates found below Earth's surface. Place the stacks of felt side by side on the table.

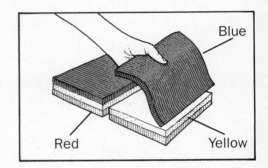

3. Push the stacks together. How is this movement like the movement of the plates below the surface? What happens to the plates? What type of mountain did you make? How do you know?

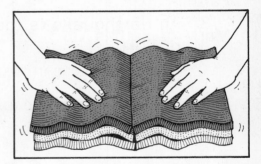

4. Draw and label the type of mountain you made.

5. Flatten the felt into two smooth stacks. Make your stacks like you did in step 2. Repeat the steps three more times.

Fold Mountains

Write the Answer
Describe how a fold mountain is made.

Two plates below the surface collide head-on

when they move. The surface folds

as the plates push up.

Fill in the circle in front of the word or phrase that best completes each sentence. The first one is done for you.

1. Plates are carried along on melted rock of the
   - ⓐ inner core.
   - ⓑ outer core.
   - ⬤ mantle.

2. Most mountains are formed in groups called
   - ⓐ peaks.
   - ⓑ ranges.
   - ⓒ plains.

3. An earthquake is the shaking of Earth's
   - ⓐ surface.
   - ⓑ inner core.
   - ⓒ outer core.

4. When the plates of Earth collide, they can form
   - ⓐ volcanoes.
   - ⓑ plains.
   - ⓒ rivers.

5. Liquid rock and hot gases blast from
   - ⓐ a hot spring.
   - ⓑ a volcano.
   - ⓒ an earthquake.

6. Heat energy from inside Earth is called
   - ⓐ solar energy.
   - ⓑ tsunami.
   - ⓒ geothermal energy.

Fill in the missing words.

7. The crust is Earth's _____outer_____ layer. (thickest, outer)

8. The crust is made up of about 20 _____plates_____. (plates, peaks)

9. Earth is made up of _____four_____ layers. (two, four)

Write the answer on the lines.

10. How can mountains be formed?

   Mountains can be formed when plates

   that make up Earth's crust move.

UNIT 7
Materials of Earth

Sulfur

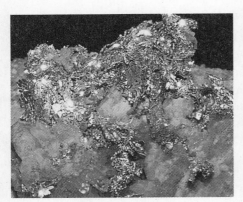
Gold in Quartz

Mica

Minerals

Minerals are solid substances formed in nature. Minerals are made of materials that were never alive. A mineral always has the same chemical make-up. The atoms of a mineral are arranged in regular patterns and form shapes called **crystals.**

Most minerals are formed by magma, the hot liquid rock inside Earth. When magma cools slowly, minerals with large crystals are formed. When magma cools quickly, minerals with small crystals form.

There are about 3,000 different minerals on Earth. Some minerals, like sulfur, are made up of one kind of atom. Others, like quartz, are made of more than one kind of atom.

Minerals have properties that can be used to identify them. One of these properties is color. Sulfur is always yellow, but other minerals, like quartz, may be found in many colors. **Luster, cleavage,** and **hardness** are other properties used to identify minerals.

Luster is how shiny a mineral is when light hits it. Some minerals, like sulfur, are dull. Others, like gold, are very shiny.

Cleavage is the way that a mineral breaks. Mica, for example, breaks in one direction and forms sheets. But sulfur breaks into many uneven pieces.

Hardness is the ability of a mineral to resist being scratched. Some minerals are so soft that they can be scratched by a fingernail. Diamond, the hardest mineral, can scratch any other material.

A. Answer True or False.

1. Minerals are liquid substances formed in nature. __False__

2. Some minerals are made up of one kind of atom. __True__

3. Most minerals are formed by earthquakes. __False__

4. The atoms of a mineral are arranged in regular patterns and form shapes called crystals. __True__

5. Minerals are made of materials that were once alive. __False__

B. Fill in the missing words.

1. A mineral always has the same __chemical__ make-up. (chemical, liquid)

2. There are about __3,000__ different minerals on Earth. (300, 3,000)

3. Minerals have __properties__ that can be used to identify them. (properties, names)

4. When magma cools __slowly__, minerals with large crystals are formed. (slowly, quickly)

5. When magma cools quickly, minerals with __small__ crystals form. (small, large)

C. Use each word to write a sentence about minerals.

Sentences will vary.

1. color _____

_____

2. luster _____

_____

3. cleavage _____

_____

4. hardness _____

_____

111

Igneous Rocks

Granite

Pumice

Obsidian

There are many kinds of rock in Earth's crust. These rocks are made up of different minerals.

Rocks can be grouped by the way they form. **Igneous rocks** form when hot liquid rock cools and hardens. This process may take place inside Earth or on its surface. Inside Earth, magma cools so slowly that it may take a thousand years for a rock to form.

One of the most common kinds of igneous rocks is granite. Granite is made by magma cooling inside Earth. Granite is light colored with large crystals. It is made up mostly of the minerals quartz and feldspar. Granite is strong and does not break easily, even under great pressure. It is often used in buildings and bridges.

Basalt is made when lava pours onto the ground or onto the ocean floor. Because the lava cools quickly, the crystals in basalt are small. When lava erupts underwater it makes pillow-shaped lumps, called pillow lava.

Sometimes, lava cools so quickly that crystals never have a chance to form. Pumice and obsidian are igneous rocks that do not have crystals. Pumice is full of bubbles and is so light that it floats. Obsidian is a glassy, black rock.

A. Answer True or False.

1. There are only a few kinds of rock in the Earth's crust. __False__

2. Rocks can be grouped by the way they form. __True__

3. Igneous rocks are formed when hot liquid rock cools and hardens. __True__

4. Basalt is made when lava pours onto the ground or onto the ocean floor. __True__

5. Sometimes lava cools so quickly that crystals never have a chance to form. __True__

B. Write the letter for the correct answer.

1. Rocks are made up of different __b__.
   (a) granite   (b) minerals   (c) lava

2. Inside Earth, magma cools so slowly that it may take a __c__ for a rock to form.
   (a) month   (b) year   (c) thousand years

3. Granite is made by magma cooling __b__ Earth.
   (a) outside   (b) inside   (c) above

4. Pumice is so light that it __c__.
   (a) fizzes   (b) erupts   (c) floats

5. Granite is light colored with __a__ crystals.
   (a) large   (b) small   (c) no

C. Draw lines to match each term with its description.

1. obsidian — used to make buildings and bridges

2. granite — glassy, black rock

3. basalt — so light it floats

4. pumice — small crystals

113

Sedimentary Rocks

Limestone

Sediments are pieces of earth or rock that have been deposited by water, wind, or ice. **Sedimentary rock** is formed from many layers of sediment piled on top of each other. It takes millions of years for sedimentary rock to form. The lowest layers of sedimentary rock are the oldest. The top layers are the newest.

This process begins when older rocks are broken down into small pieces by weathering. The pieces of rock may be picked up and moved by rivers. When the water slows down, the large, heavy sediments are the first to settle. Lighter sediments fall on top of the heavier ones.

Some sediments are deposited in lakes and oceans. Over many years, the layers of sediments build up. The top layers press down on the lower layers. This pressure slowly turns the bottom layers into rock. These layers of rock are called **strata.** Strata of shale and limestone are formed in this way.

Some sedimentary rocks are formed when remains of living things are deposited in layers. For example, chalk is formed when shells from small sea animals are cemented together with minerals.

Some sedimentary rocks form when water evaporates, or goes into the air, and leaves behind deposits of minerals. When sea water evaporates, rock salt or gypsum may be formed.

A. Answer True or False.

1. Sedimentary rock is formed from many layers of sediment.
   ___True___

2. When rocks are broken down, their particles always stay in one place. ___False___

3. Some sedimentary rocks are formed from the remains of living things. ___True___

4. Some sedimentary rock is formed when it rains. ___False___

B. Use the words below to complete the sentences.

chalk	Sediments	strata
gypsum	shale	weathering

1. Layers of rock are called ___strata___.

2. Strata of ___shale___ and limestone are made when the top layers of sediments press down on the lower layers, turning them to rock.

3. Older rocks are broken down into small pieces by ___weathering___.

4. When sea water evaporates, rock salt or ___gypsum___ may be formed.

5. ___Sediments___ are pieces of earth or rock that have been deposited by water, wind, or ice.

C. Write the letter for the correct answer.

1. When water slows down, ___c___, heavy sediments are the first to settle.
   (a) small   (b) tiny   (c) large

2. Some sediments are deposited in lakes and ___a___.
   (a) oceans   (b) mountain peaks   (c) ridges

3. The lowest layers of sedimentary rock are the ___c___.
   (a) wettest   (b) newest   (c) oldest

Metamorphic Rocks

Pressure changes shale into slate.

Slate

Metamorphic rocks are made when igneous and sedimentary rocks are changed into a different type of rock. Sometimes, even metamorphic rocks are changed into a different kind of rock.

This happens when the weight of many layers pushes rock deep inside Earth. Here heat and pressure from the layers above change the form of these rocks.

The liquid rock, or magma, inside Earth heats the rock to a very high temperature. Some minerals change shape, and others break down and react with each other to form new minerals.

Rubies and sapphires are some of the minerals that are formed by high temperature. They are formed when magma comes into contact with limestone.

Pressure from movements of Earth's crust can change rock. This often happens during the formation of mountains. Pressure can change shale to slate. Emeralds are formed within metamorphic rocks as a result of great pressure.

The rocks that make up Earth are always changing. Igneous, sedimentary, and metamorphic rocks can be broken into sediments and deposited into layers to form new sedimentary rocks.

All of these kinds of rock can be changed by heat or pressure into new metamorphic rocks. If rock becomes buried deeply enough, it melts back into magma. When the magma cools, new igneous rocks are formed. This constant changing of rocks is called the **rock cycle.**

A. Answer True or False.

1. The rocks that make up Earth are always changing. __True__

2. Metamorphic rocks can be formed only from sedimentary rocks. __False__

3. Rocks may be changed by pressure. __True__

4. The magma inside Earth heats the rock to a very high temperature. __True__

5. The constant changing of rocks is called the rock cycle. __True__

6. When magma heats rock to a very high temperature, some minerals change shape. __True__

7. Metamorphic rocks are formed by glaciers. __False__

B. Fill in the missing words.

1. Metamorphic rocks are made when igneous, sedimentary, and __metamorphic rocks__ are changed into a different type of rock. (metamorphic rocks, water)

2. Rocks may be changed by __heat__ and pressure. (cold, heat)

3. The constant changing of rocks is called the __rock cycle__. (water cycle, rock cycle)

4. When magma cools, new __igneous__ rocks are formed. (igneous, sedimentary)

5. Pressure from movements of Earth's __crust__ can change rock. (core, crust)

C. Draw lines to match each rock and how it was formed.

1. rubies   used to be shale

2. slate   formed by high temperatures

3. emeralds ———— formed as a result of great pressure

Ores

Sulfur

Some rocks are called **ores.** Ores are minerals or rocks with enough metals in them to make them worth mining. Most ores are found in beds with rocks of little value. Ore beds close to the surface are easy to mine. But others are deep within the ground.

Some ores contain metals like iron and copper. Metals are usually shiny substances. They can transfer heat and electricity. Some metals can be pounded flat or pulled into wires without breaking. Lead, gold, and aluminum are metals found in ores.

Metal ores are important because so many products are made from them. From gold, we make jewelry and fillings for teeth. Lead is used to make pipe, and copper is used to make wire. Aluminum is used to make airplanes, cans, foil, cooking pots, and motors. Iron is used to make steel.

Some ores contain minerals like sulfur and graphite. Sulfur is important because it is used in making medicines and fertilizers. Graphite, a form of carbon, is the "lead" in most pencils.

Underline the correct words.

1. Many (products, minerals) are made from metal ores.

2. Some ores contain (metals, medicines) like iron and copper.

Fossils

Fossils are the remains or imprints of plants and animals that died long ago. Fossils are usually found in sedimentary rock.

When a dead animal or plant gets covered by sediment, a **mold** or **cast** of the fossil may form. As the sediments harden into rock, the soft parts of the dead matter will decay. The empty space left in the rock is called a mold. It has the same shape as the plant or animal. If the mold fills with minerals from water, the minerals may harden to form a cast. The cast also has the same shape as the living thing that died.

Traces of animals, such as footprints, may also become fossils. Footprints are often found in mud that later hardened into rock.

Whole plants or animals are the rarest kind of fossil. But entire giant wooly mammoths have been found frozen in ice.

Fossils are an important source of information about the past. For example, if scientists find fossils of shells on dry land, they know that that part of Earth was once covered with water.

Cast and Mold Fossils

Dinosaur Footprints

Underline the correct words.

1. (Fossils, Rocks) are the remains or imprints of plants and animals that died long ago.

2. Fossils are usually found in (sedimentary, igneous) rock.

3. Footprints of animals (may, may not) become fossils.

4. Giant wooly mammoths have been found frozen in (mud, ice).

Coal

Mining Coal

Anthracite and Bituminous Coal

Coal is a solid fossil fuel. It is a rock made from decayed plants that lived long ago. It takes millions of years for dead plants to change to coal. The three major forms of coal are **lignite, bituminous coal,** and **anthracite.** Each form of coal can be used as a fuel.

Millions of years ago, huge ferns grew in the swamps on Earth. When the plants died, they fell into the water. Layer after layer of plants gathered. The weight of the top layers pressed down on the bottom layers. Over the years, the pressure hardened the plants, forming a substance called **peat.** Peat can be used as fertilizer or burned like coal. But it does not burn as hot or as completely as real coal.

As more time passed, sediment such as sand piled on top of the peat. Slowly, the sand turned into sandstone and other types of rock. Under this pressure, the peat changed to lignite. Lignite is dark brown and burns easily. More pressure on lignite formed bituminous, or soft coal. Bituminous coal is black and gives off a lot of heat when it is burned. This makes it a very good fuel.

Anthracite is formed when bituminous coal is under great pressure for many years. Anthracite coal is black, hard, and shiny.

Coal is mined because it is an important fuel. It is also used to make medicines, dyes, and fertilizers. Coal is used to produce about two thirds of the world's electricity.

120

A. Answer True or False.

1. Coal is a solid fossil fuel. _____True_____

2. Coal is mined because it is an important fuel. _____True_____

3. Peat cannot be burned. _____False_____

4. The three major forms of coal are lignite, bituminous coal, and anthracite. _____True_____

5. Coal is a rock made from decayed animals. _____False_____

6. Anthracite coal is black, hard, and shiny. _____True_____

7. Coal is used to make medicines, dyes, and fertilizers. _____True_____

B. Draw lines to match each term with its description.

1. peat   soft, black, and gives off a lot of heat, which makes it a good fuel

2. lignite   does not burn as hot or as completely as real coal

3. bituminous   dark brown and burns easily

4. anthracite ——————— black, hard, and shiny

C. The sentences tell how coal is formed. Put these sentences in the correct order. The first one is done for you.

_____5_____ More pressure changed peat to lignite.

_____3_____ Layer after layer of plants gathered.

_____1_____ Millions of years ago huge ferns grew in swamps.

_____4_____ Pressure hardened the plants, forming peat.

_____6_____ Pressure on lignite forms bituminous coal.

_____7_____ Anthracite is formed when bituminous coal is under pressure.

_____2_____ When the plants died, they fell into the water.

Petroleum

Drilling for Oil in the Ocean

Drilling for Oil on Land

Petroleum is a liquid fossil fuel. It formed from tiny plants and animals that lived in the ocean long ago. Much more of Earth's surface was covered with water then. When these plants and animals died, they dropped to the bottom of the ocean. Layers of mud drifted over them. After many years, the dead matter was pressed down into sedimentary rocks. This pressure changed the materials into oil.

Petroleum is an important fossil fuel because it has so many uses. It is made into gasoline for automobiles, trucks, trains, airplanes, and ships. It is used to produce electricity for houses and businesses. Thousands of products, like asphalt, fabrics, carpets, detergents, and plastics, are made from petroleum.

Nearly half the energy used in the world comes from petroleum. We are rapidly using up the world's supply of this important resource. As wells on land run dry, new wells are being drilled in the ocean floor. Drilling wells in the ocean is more expensive and dangerous than drilling on land. But scientists hope to find more petroleum there.

Underline the correct words.

1. Petroleum is a (solid, liquid) fossil fuel.

2. Petroleum is formed from tiny plants and animals that lived in the (ocean, ground) long ago.

3. New wells are being drilled in the (ocean floor, desert).

4. Petroleum is made into gasoline and (coal, plastics).

Part A

Use the words below to complete the sentences.

coal	igneous	ores
crystals	metamorphic	petroleum
Fossils	mineral	sedimentary

1. A _____mineral_____ always has the same chemical make-up.

2. When hot liquid rock cools and hardens, _____igneous_____ rocks are formed.

3. Rocks that are formed from sediments are called _____sedimentary_____ .

4. Rocks with enough metals to make them worth mining are _____ores_____ .

5. Rocks that have been changed into new types of rocks are _____metamorphic_____ rocks.

6. _____Fossils_____ are the remains or imprints of plants and animals that died long ago.

7. A solid fossil fuel made from plants that lived long ago is _____coal_____ .

8. Tiny ocean plants and animals formed a liquid fossil fuel when pressure changed them to _____petroleum_____ .

9. The atoms of a mineral are arranged in regular patterns and form shapes called _____crystals_____ .

Part B

Draw lines to match the terms with their descriptions.

1. luster   provides half the world's energy

2. rock cycle   how shiny a mineral is

3. petroleum   changes into coal under pressure

4. peat   the constant changing of rocks

Make a Fossil

You Need
- ½ stick clay
- 1 small shell
- 1 small rock
- 1 leaf

1. Roll the clay into a ball. Then flatten it with your hand.

2. Choose a small shell, a rock, or a leaf. Press it into the clay. Carefully pull it out of the clay. What do you see? How is your imprint like the imprint of a fossil?

3. Divide the clay into two pieces. Roll each piece into a ball. Flatten both balls.

4. Press a shell, a rock, or a leaf into one piece of clay. Place the other piece of clay over the shell, the rock, or the leaf. How are these clay layers like the sedimentary rock that contains fossils?

5. Carefully take off the top layer of clay. Pull the shell, the rock, or the leaf out of the clay. Look at each piece of clay. What does each piece of clay look like?

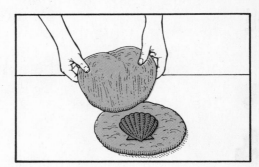

Write the Answer
Describe how a fossil is made.

A plant or a dead animal is covered by sediment. The sediment hardens into rock.

The plant or animal decays and is washed away leaving a space shaped like the

plant or the animal. Minerals fill the shape and harden to form a fossil.

Fill in the circle in front of the word or phrase that best completes each sentence. The first one is done for you.

1. Igneous rocks are made when hot liquid rock
   - ⓐ heats up and erupts.
   - ● cools and hardens.
   - ⓒ moves and swirls.

2. Layers of sediments form
   - ⓐ igneous rocks.
   - ⓑ metamorphic rocks.
   - ⓒ sedimentary rocks.

3. Rocks that have been changed into new types of rocks are
   - ⓐ igneous rocks.
   - ⓑ metamorphic rocks.
   - ⓒ sedimentary rocks.

4. Coal is a
   - ⓐ solid fossil fuel.
   - ⓑ liquid fossil fuel.
   - ⓒ hot liquid rock.

5. Fossils are the remains or imprints of plants and animals formed in
   - ⓐ sedimentary rock.
   - ⓑ water.
   - ⓒ air.

6. The constant changing of rocks is called
   - ⓐ the heat cycle.
   - ⓑ strata.
   - ⓒ the rock cycle.

Fill in the missing words.

7. Most minerals are formed by _____magma_____. (magma, lava)

8. Some ores are mined for their _____metals_____. (metals, fossils)

9. Gasoline is made from _____petroleum_____. (coal, petroleum)

Write the answer on the lines.

10. What are two ways that minerals can be identified?

   _____Answers will vary_____

   _____but should include color, cleavage, hardness, or luster._____

UNIT 8
Conservation

Plants and animals are renewable resources.

Fossil fuels, such as the oil being drilled from this well, are nonrenewable resources.

Earth in Balance

Natural resources are things found in nature that people use. When left alone, the resources of Earth stay in natural balance. Materials are used, then replaced.

The balance of natural resources affects all living things. For example, sulfur is a mineral found in soil. Plants use sulfur as they make food. Animals eat the plants and use the sulfur to grow. When plants and animals die, their bodies decay and the sulfur goes back into the soil. In this way, sulfur is replaced, and this natural resource stays in balance.

Natural resources that can replace, or renew, themselves in a short time are called **renewable natural resources.** Plants and animals are renewable natural resources, because they produce offspring, or new living things.

Natural resources that are replaced very slowly or not at all are called **nonrenewable natural resources.** Soil is a nonrenewable resource, because it takes many years for rocks to break down and become soil. Nonrenewable resources are not living things. They include fossil fuels and minerals.

Conservation is the protection of Earth's natural balance and the wise use of its resources. Conservationists are people who make sure natural resources are not destroyed. They identify plants and animals in danger of dying out and try to save them. Conservationists try to find ways to use fuels, like coal, wisely. They also try to find other, safer fuels.

Answer <u>True</u> or <u>False</u>.

1. Natural resources are things found in nature that people use.

   _____True_____

2. When left alone, the resources of Earth stay in natural balance.

   _____True_____

3. Nonrenewable natural resources replace themselves in a short

   time. _____False_____

4. Conservationists are people who make sure natural resources are

   not destroyed. _____True_____

B. **The sentences below tell how sulfur stays in a natural balance. Put the sentences in the correct order. The first one is done for you.**

_____2_____ Plants use sulfur as they make food.

_____3_____ Animals eat the plants and use sulfur to grow.

_____1_____ Sulfur is a mineral found in soil.

_____4_____ When plants and animals die, their bodies decay and the sulfur goes back into the soil.

_____5_____ In this way, sulfur is replaced and stays in balance.

C. **Fill in the missing words.**

1. Plants and animals are _____renewable_____ resources. (renewable, nonrenewable)

2. Plants and animals produce offspring, or new

   _____living things_____. (elements, living things)

3. Nonrenewable natural resources are materials such as

   _____soil_____. (living things, soil)

4. Conservation is the protection and _____wise_____ use of Earth's resources. (careless, wise)

5. Conservationists try to find ways to use _____fuels_____ wisely. (fuels, people)

Saving the Soil

Contoured Farmland

Soil is one of Earth's most important resources. Most of our food comes from plants that grow in soil or from the animals that eat these plants.

One way to save, or conserve, soil is to keep it rich in **nutrients.** Nutrients are the materials that plants need to grow. **Crop rotation** helps keep soil healthy. Farmers rotate crops by planting one crop the first season and another crop the following season. For example, corn is planted one season. Corn uses up nitrogen in the soil. The next season, soybeans are planted. Soybeans put nitrogen back into the soil.

Another way to conserve soil is to keep it from being carried away. Farmers can prevent soil from being carried away by using **strip cropping** and **contour plowing.** When farmers use strip cropping, they plant strips of different crops across a field. The first strip may be grass. The second strip may be wheat. Because grass holds soil in place better than wheat, it keeps soil from being carried away by wind or water.

In contour plowing, rows are plowed around hills instead of up and down them. This slows down run-off. Remember that run-off is water that flows over soaked ground. Run-off flows quickly downhill but is slowed by the plowed rows. This saves soil because slower-moving water carries away less soil than fast-moving water.

Land may also be reclaimed when soil is put back on the surface of Earth. Sometimes, mining companies reclaim land after they have dug holes to mine for coal.

A. Answer True or False.

1. Most of our food comes from plants that grow in soil or from the animals that eat these plants. _____True_____

2. Crop rotation helps keep soil healthy. _____True_____

3. Strip cropping is a way to carry soil away. _____False_____

4. In contour plowing, rows are plowed up and down hills. _____False_____

5. Land is reclaimed when soil is put back on the surface of Earth.
   _____True_____

6. Soil is not an important resource. _____False_____

7. One way to conserve soil is to keep it rich in nutrients. _____True_____

B. Fill in the missing words.

1. Corn _____uses up_____ nitrogen in the soil. (uses up, puts)

2. In contour plowing, rows are plowed _____around_____ hills. (around, up and down)

3. Because grass holds soil in place better than wheat, it keeps soil from being _____carried away_____. (plowed, carried away)

4. Run-off is _____slowed down_____ by contour plowing. (slowed down, speeded up)

5. Sometimes, mining companies _____reclaim_____ land after they have dug holes to mine for coal. (reclaim, plow)

C. Use each word to write a sentence about saving the soil.

1. crop rotation _____Sentences will vary._____

   _____

2. strip cropping _____

   _____

3. contour plowing _____

   _____

Keeping Air Clean

Factory smokestacks send pollutants into the air.

People's activities often upset the natural balance of Earth. **Air pollution,** or dirty air, is a serious problem caused by people.

Most air pollution is caused by burning fossil fuels. Using gasoline in cars and burning coal to make electricity are major causes of air pollution. Cars and factories send harmful materials, called pollutants, into the air. Factory smokestacks also send pollutants into the air.

The hazy pollution that often hangs over cities is called **smog.** Cars, factories, and furnaces pour tons of pollutants into cities each day. Smog makes breathing difficult. It can cause serious diseases.

When air pollution mixes with moisture in the air, it makes strong chemicals called acids. These acids fall to Earth when it rains. Acid rain is dangerous. When acids pollute lakes, fish die, and the water is not safe to drink.

We can reduce the amount of pollutants going into the air. If people ride together in car-pools, fewer cars will be on the road. Using buses or trains also reduces the number of cars used. This means less fuel will turn into air pollution.

People can insulate, or put special materials into the walls of their homes. Then less energy is needed to heat or cool them. Factories can use special screens so pollutants are not sent into the air.

A. Answer True or False.

1. Air pollution is the cleaning of the air. __False__

2. Most air pollution is caused by burning fossil fuels. __True__

3. The hazy pollution that hangs over cities is called fog. __False__

4. Smog makes breathing easier. __False__

5. Acid rain can pollute lakes. __True__

6. Factory smokestacks send pollutants into the air. __True__

B. Fill in the missing words.

1. Burning __coal__ to make electricity is one cause of air pollution. (acids, coal)

2. People's activities often __upset__ the natural balance of Earth. (upset, protect)

3. When air pollution mixes with moisture, strong chemicals called __acids__ are made. (acids, sprays)

4. Using buses or trains reduces the number of __cars__ used. (cars, smokestacks)

5. People can insulate their homes so less __energy__ is needed. (air, energy)

6. Factories can use special screens so __pollutants__ are not sent into the air. (pollutants, water)

7. When acids pollute lakes, fish die, and the water __is not__ safe to drink. (is, is not)

C. Draw lines to match each term with its effect.

1. acid rain   causes serious diseases

2. smog   pollutes lakes

3. fossil fuels   reduce the number of cars on the road

4. car-pools   cause air pollution when burned

131

Keeping Water Clean

Trash and other human waste products pollute water.

Oil spills can cause great damage.

Water is one of Earth's most important resources, because all plants and animals need water to live. In the United States, each person uses about 70 gallons of water a day. We use water for drinking, cooking, bathing, watering crops, and producing electricity.

The main cause of water pollution is the release of harmful materials by factories. Some factories dump strong chemicals into streams and rivers. Other factories release heated water. Chemicals and hot water can kill plant and animal life in streams and lakes.

Sometimes, cities dump raw sewage, or human wastes, into rivers. Run-off may also carry harmful wastes. Fertilizers and pesticides used on farms may be carried into waterways. Polluted water tastes bad, kills fish, and makes people sick.

Some oceans are polluted by oil. Sometimes, oil leaks from the rigs that pump petroleum up from the ocean floor. Other times, oil spills from huge ships called tankers. Tankers carry millions of gallons of oil. A tanker accident can cause great damage. An oil spill makes beaches dirty and kills birds, fish, and other sea animals.

There are many ways we can conserve water. Special water-saving devices can be put in toilets and showers. We can make sure our faucets do not leak. Stricter laws can be passed that keep factories from dumping harmful materials into water. Laws can also force all cities to have proper sewage treatment plants.

A. Answer True or False.

1. Living things do not need water. _____False_____

2. The main cause of water pollution is the release of harmful materials by factories. _____True_____

3. Chemicals and hot water can kill plant and animal life in streams and lakes. _____True_____

4. Run-off may also carry harmful wastes. _____True_____

5. There are many ways we can conserve water. _____True_____

B. Write the letter for the correct answer.

1. In the United States, each person uses about _____b_____ gallons of water a day.
   (a) 7   (b) 70   (c) 700

2. We use water for drinking, cooking, bathing, watering crops, and producing _____c_____ .
   (a) wastes   (b) coal   (c) electricity

3. Sometimes, cities dump raw sewage, or _____a_____ , into rivers.
   (a) human wastes   (b) air   (c) clean water

4. Fertilizers used on farms may be carried into _____a_____ .
   (a) waterways   (b) mountains   (c) air

5. Some oceans are polluted by _____c_____ .
   (a) gas   (b) minerals   (c) oil

C. Fill in the missing words.

1. Polluted water tastes _____bad_____ and makes people sick. (good, bad)

2. Sometimes, oil that pollutes the ocean comes from _____rigs_____ . (airplanes, rigs)

3. Tanker accidents can cause great _____damage_____ . (damage, benefits)

4. An oil _____spill_____ makes beaches dirty. (pump, spill)

Saving Our Resources

These deer are protected in a wildlife refuge.

For many years, people were not concerned with conservation. They thought natural resources would last forever.

Recently, however, much has been written about air pollution and water pollution. People know that we cannot take clean air and water for granted.

People make too much garbage. We are running out of places to put it. Many cities now have recycling. People separate things that can be reused. Paper, aluminum cans, and glass are some things we can recycle. There are special places to take old motor oil so it won't end up in the soil and water.

Many plants and animals in the world are in danger of dying out. We call these plants and animals **endangered species.** We make places to keep them safe. These places are called **wildlife refuges.** These refuges have saved animals, such as the whooping crane, pelican, and bison.

Underline the correct words.

1. For years, people were not concerned with (laws, <u>conservation</u>).

2. (<u>Endangered species</u>, Environments) are plants and animals in danger of dying out.

3. Wildlife refuges are places where plants and animals are (hunted, <u>kept safe</u>).

Part A

Fill in the missing words.

1. Natural resources are things found in nature that people ___**use**___.
   (use, find)

2. Conservation is the wise use of Earth's ___**resources**___.
   (animals, resources)

3. Farmers can use strip cropping to keep soil from being
   ___**carried away**___. (held in place, carried away)

4. Most air pollution is caused by burning fossil ___**fuels**___.
   (fuels, forests)

5. The main cause of water pollution is the release of harmful materials
   by ___**factories**___. (forests, factories)

6. Many plants and ___**animals**___ in the world are in danger of
   dying out. (animals, people)

7. Polluted water kills fish and makes people ___**sick**___. (well, sick)

Part B

Underline the correct words.

1. Tankers are huge (ships, airplanes) that carry oil.

2. Nonrenewable natural resources are resources that are replaced
   (in a short time, very slowly).

3. Crop rotation is a way (miners, farmers) keep soil rich in nutrients.

4. When farmers use (strip cropping, crop rotation), they plant strips of
   different crops across a field.

5. Contour plowing slows (forest fires, run-off).

6. Acid rain is a mix of air pollution and (dirt, moisture).

7. When left alone, the resources of Earth stay in natural
   (balance, pollution).

Make a Conservation Poster

You Need

● **2 partners** ● **11 x 18 construction paper** ● **markers**

1. Review the word *conservation* on page 126. How does a conservationist help plants and animals? How does a conservationist help people?

2. Make a list and talk with your group about ways your community could conserve its natural resources better. Does your community have a problem with air pollution or poor soil? Does your community recycle? What do you think could be done to protect these resources?

3. Choose one problem and a solution your group would like to tell the class about. What could you say or draw on a poster?

4. Work together to draw the poster.

5. Show the class your conservation poster. Explain what you think the problem is and how your group wants to correct it.

Write the Answer
Why do people need to take care of the air and the water?

Water and air are being polluted.

People can drink polluted water or breathe polluted

air and get very sick.

Fill in the circle in front of the word or phrase that best completes each sentence. The first one is done for you.

1. Strip cropping and contour plowing help keep soil
   a. rich in materials.
   b. moving.
   ● from being carried away.

2. Burning fossil fuels is the cause of most
   ● air pollution.
   b. soil erosion.
   c. endangered species.

3. People can help keep air clean by
   ● riding in car-pools.
   b. burning fuels.
   c. putting soot in the air.

4. A wildlife refuge is a place where plants and animals are
   a. hunted.
   b. endangered.
   ● kept safe.

5. Some factories pollute water by
   a. using screens to keep soot out of the air.
   ● dumping chemicals into streams.
   c. planting crops.

6. Every living thing needs
   a. fuels.
   b. pollution.
   ● water.

Fill in the missing words.

7. Crop rotation is a way to keep soil rich in _____nutrients_____. (water, nutrients)

8. Materials are used and _____replaced_____ in Earth's natural balance. (wasted, replaced)

9. Some things we can _____recycle_____ are paper, aluminum cans, and glass. (discard, recycle)

Write the answer on the lines.

10. What is a renewable natural resource?

   A renewable natural resource
   _____
   is a natural resource that can replace itself in a short time.

Glossary

A **abyssal plain,** page 56.
The abyssal plain is a wide, flat area that makes up most of the ocean floor.

Africa, page 60.
Africa is the second largest of the seven continents. It is connected to Asia.

air mass, page 36.
An air mass is a huge body of air that stays together as it moves.

air pollution, page 130.
Air pollution is dirty air. It is caused mostly by burning fossil fuels.

air pressure, page 38.
Air pressure is the push of air on Earth.

altitude, page 76.
Altitude is the height of an area above sea level.

Antarctica, page 60.
Antarctica is the coldest of the seven continents. The South Pole is in Antarctica.

anthracite, page 120.
Anthracite is a form of coal produced when bituminous coal is under great pressure for many years.

Arctic, page 78.
The Arctic is the area around the North Pole. It includes the northern parts of Europe, Asia, and North America.

Asia, page 60.
Asia is the largest of the seven continents. It is connected to Europe and Africa.

asteroid, page 4.
An asteroid is an object made of rock and metal. Asteroids orbit the sun.

atmosphere, page 10.
The atmosphere is a special mixture of gases around Earth.

Australia, page 60.
Australia is the smallest of the seven continents. It has water all around it.

axis, page 21.
The axis of Earth is an imaginary line that goes through Earth from the North Pole to the South Pole.

B **banks,** page 46.
The banks are the sides of a river.

barometer, page 38.
A barometer is an instrument that measures air pressure.

basin, page 48.
A basin is the low ground that holds a lake.

bed, page 46.
The bed is the bottom of a river.

bedrock, page 90.
Bedrock is solid rock under layers of soil.

bituminous coal, page 120.
Bituminous, or soft, coal is formed when lignite is under pressure.

C **canyon,** page 62.
A canyon is a deep valley with steep sides.

cast, page 119.
A cast is a type of fossil. It is formed when a mold has filled with minerals that harden into rock.

cave, page 52.
A cave is an underground hole big enough for a person to enter.

cinder cone, page 104.
A cinder cone is a cone-shaped mountain made when a volcano erupts violently and throws cinders out the top.

cinders, page 104.
Cinders are volcanic rocks about the size of golf balls.

cirrus, page 30.
Cirrus clouds are white and feathery. They form high in the sky and are made of ice crystals.

clay, page 90.
Clay is soil made of mica and other minerals. It has tiny grains that can be seen only with a microscope.

cleavage, page 110.
Cleavage is the way that a mineral breaks.

climate, page 68.
Climate is the average weather of a region over a long period of time.

coal, page 120.
Coal is a solid fossil fuel made from decayed plants that lived long ago.

column, page 52.
A column is a tall, thin rock that forms when a stalactite and a stalagmite meet.

comet, page 4.
A comet is an object made of ice, gas, and dust. Comets orbit the sun.

composite volcano, page 104.
A composite volcano is formed by layers of cinders and layers of lava flows.

condensation, page 28.
Condensation is the process by which water vapor in the air changes to liquid water.

conservation, page 126.
Conservation is the protection of Earth's natural balance and the wise use of its resources.

constellation, page 8.
A constellation is a group of stars that form a pattern.

continents, page 60.
Continents are the seven largest land masses on Earth.

contour plowing, page 128.
Contour plowing is a way farmers prevent soil erosion. Rows are plowed around hills instead of up and down, to slow run-off.

crop rotation, page 128.
Crop rotation is a way to keep soil rich. One crop is planted the first season, and another crop in the following season.

crust, page 96.
The crust is the outer layer of Earth. It makes up the continents and ocean floors.

crystal, page 110.
A crystal is a solid in which the atoms are arranged in regular patterns.

cumulus, page 30.
Cumulus clouds are thick, white, and fluffy. They are made of water droplets.

current, page 68.
A current is a moving stream of water in an ocean.

D **deciduous forest,** page 72.
A deciduous forest grows where winters are cold and summers are warm. Deciduous trees lose all their leaves in the fall.

degree, page 35.
A degree is a unit for measuring temperature.

desert, page 70.
A desert is a place that gets less than 10 inches of rain in a year.

dew, page 32.
Dew is water vapor that condenses and forms drops of water as the air cools.

diameter, page 6.
Diameter is the distance through a circular object from one side to the other.

dome-shaped mountains, page 100.
Dome-shaped mountains are formed when two plates push against each other and one plate bulges out.

dune, page 70.
A dune is a tall pile of sand that is continually moved and shaped by the wind.

E **earthquake,** page 102.
An earthquake is a shaking of the surface of Earth when the plates of Earth move.

endangered species, page 134.
Endangered species are kinds of plants and animals that are in danger of dying out.

epicenter, page 102.
The epicenter is the place on the surface of Earth directly above the focus of an earthquake.

equator, page 8.
The equator is the imaginary line around the center of Earth.

erosion, page 82.
Erosion is the wearing down and moving of rocks and soil from one place to another.

erupt, page 104.
To erupt is to burst out, or explode. For example, a mixture of liquid rock and hot gases may erupt from a volcano.

Europe, page 60.
Europe is one of the seven continents. It is connected to the continent of Asia.

evaporation, page 28.
Evaporation is the process by which water changes from a liquid to a gas and goes into the air.

evergreen, page 72.
An evergreen tree is a tree that stays green all year long.

F **fault,** page 98.
A fault is a line where two plates of Earth's crust pass each other.

fault-block mountains, page 100.
Fault-block mountains are formed by blocks of rock that are pushed up along fault lines. The rock is broken when plates collide.

flowing artesian well, page 50.
A flowing artesian well is a well that can produce water without pumping.

focus, page 102.
The focus is the point where rocks in the crust break and move apart during an earthquake.

fog, page 30.
Fog is a stratus cloud that forms on the ground.

fold mountains, page 100.
Fold mountains are formed when plates collide head-on, folding the crust and forming peaks.

forest, page 72.
A forest is a place where many trees grow.

fossil, page 119.
A fossil is the remains or the imprint of a plant or animal that died long ago.

fresh water, page 48.
Fresh water is water that contains less dissolved minerals than salt water.

front, page 36.
A front is a boundary that forms when two air masses meet.

frost, page 32.
Frost is dew that has frozen.

G geothermal energy, page 106.
Geothermal energy is heat inside Earth. Hot springs and geysers are sources of geothermal energy.

geyser, page 106.
A geyser is a place where hot water and steam shoot out of the ground like a volcanic eruption.

glacier, page 86.
A glacier is a large, slow-moving sheet of ice.

grain, page 74.
A grain is a grass such as wheat, corn, or oats.

grasslands, page 74.
Grasslands are areas where many kinds of grasses grow. They are found in both temperate and tropical climates.

gravity, page 4.
Gravity is a force that acts like a pull between the sun and the planets.

ground water, page 50.
Ground water is underground water. It is rain that has soaked into the soil and collected above a layer of rock.

H hardness, page 110.
Hardness is the ability of a mineral to resist being scratched.

high, page 38.
A high is an area of high pressure formed when cold air falls to the ground.

humidity, page 34.
Humidity is the amount of water vapor in the air.

humus, page 90.
Humus is the remains of dead plants and animals found in soil.

hurricane, page 40.
A hurricane is a large storm that forms over an ocean near the equator.

I igneous rock, page 112.
Igneous rock is formed when hot liquid rock cools and hardens.

inner core, page 96.
The inner core is at the center of Earth. It is 800 miles thick and is made of solid iron and nickel.

inner planets, page 10.
The inner planets are the four planets closest to the sun. Mercury, Venus, Earth, and Mars are the inner planets.

L lake, page 48.
A lake is a body of fresh water with land all around it. It is bigger than a pond.

latitude, page 68.
Latitude is the distance north or south of the equator.

lava, page 100.
Lava is hot, melted rock spewed onto the surface of Earth by a volcano.

lightning, page 40.
Lightning is a flash of light caused by electric charges in a cloud.

lignite, page 120.
Lignite is a form of coal that is made when peat is under pressure.

limestone, page 52.
Limestone is a rock found in most caves.

low, page 38.
A low is an area of low pressure formed when warm air rises.

luster, page 110.
Luster describes how shiny a mineral is when light hits it.

M mantle, page 96.
The mantle is the thickest layer of Earth. It lies just below the crust.

mesosphere, page 26.
The mesosphere is the layer of the atmosphere beyond the stratosphere.

metamorphic rock, page 116.
Metamorphic rock is formed when igneous and sedimentary rock are changed.

meteorologist, page 42.
A meteorologist is a person who studies the weather.

mineral, page 110.
A mineral is a solid substance formed in nature.

mold, page 119.
A mold is a type of fossil. It is the empty space formed when the soft parts of dead matter decay and the sediments around the dead matter harden into rock.

moon, page 4.
A moon is a small object in orbit around a planet.

moraine, page 86.
A moraine is a ridge of material left behind as a glacier melts.

mountain, page 62.
A mountain is a landform that is 2,000 feet or more above the land around it.

mouth, page 46.
The mouth is the end of a river, where the river empties into another body of water.

N **natural resources,** page 126.
Natural resources are things found in nature that people use.

nonrenewable natural resources, page 126.
Nonrenewable natural resources are those that are replaced slowly or not at all.

North America, page 60.
North America is one of the seven continents. It is connected to the continent of South America.

North Pole, page 8.
The North Pole is the northernmost point on Earth.

Northern Hemisphere, page 22.
The Northern Hemisphere is the part of Earth above the equator.

nutrients, page 128.
Nutrients are the materials that plants need to grow.

O **oasis,** page 70.
An oasis is an area in a desert that has water.

ocean, page 54.
An ocean is a large, deep body of salt water.

orbit, page 4.
An orbit is a curved path. Each of the planets moves in an orbit around the sun.

ore, page 118.
An ore is a mineral or a rock that has enough metal in it to make it worth mining.

outer core, page 96.
The outer core is a layer of Earth. It is 1,400 miles thick and is made of hot liquid metal.

outer planets, page 16.
The outer planets are the five planets farthest from the sun. Jupiter, Saturn, Uranus, Neptune, and Pluto are the outer planets.

P **peak,** page 62.
A peak is the pointed top of a mountain.

peat, page 120.
Peat is a substance formed by decayed plants under pressure for millions of years.

petroleum, page 122.
Petroleum is a liquid fossil fuel formed from tiny ocean plants and animals that lived in the ocean long ago.

phases, page 14.
Phases are the changing views of the moon you can see in a month.

plain, page 64.
A plain is a large area of flat land that is usually lower than the land around it.

planet, page 4.
A planet is a large, solid object that travels around a star.

plate, page 98.
A plate is a large section of Earth's crust. There are about 20 plates.

plateau, page 64.
A plateau is a large area of flat land that rises above the land around it.

polar, page 68.
In a polar climate zone, it is cold much of the year and little rain or snow falls.

pond, page 48.
A pond is a body of fresh water that has land all around it. A pond is smaller and shallower than a lake.

precipitation, page 32.
Precipitation is water that falls to Earth. Rain, snow, sleet, and hail are forms of precipitation.

R **range,** page 62.
A range is a group of mountains.

reflect, page 6.

To reflect means to send back. Planets and moons reflect light from the sun.

renewable natural resources, page 126.

Renewable natural resources are those that can renew, or replace, themselves in a short time.

ridge, page 62.

A ridge is a long, narrow place on a mountain where two sloping surfaces meet.

river, page 46.

A river is a natural flow of water that runs into another body of water.

rock cycle, page 116.

The rock cycle is the constant changing of rock from one kind to another.

rotation, page 21.

Rotation is a spinning movement. Earth rotates on its axis once every 24 hours.

run-off, page 46.

Run-off is water that runs across ground that cannot soak up any more water.

salt water, page 54.

Salt water is water that has many minerals, such as salt, dissolved in it. The water in the oceans is salt water.

sand, page 90.

Sand is a kind of soil made of feldspar. It has the biggest grains of any kind of soil.

sandbar, page 84.

A sandbar is a deposit of sand and soil near the shore that is made by ocean waves.

satellite, page 14.

A satellite is an object in space that orbits another object in space.

sedimentary rock, page 114.

Sedimentary rock is formed from many layers of sediment piled on top of each other.

sediments, page 114.

Sediments are pieces of earth or rock that have been deposited by water, wind, or ice.

shelf, page 56.

The shelf is the first 100 miles of the ocean floor, from the shoreline. The shelf is smooth and flat and slowly gets deeper.

shield volcano, page 104.

A shield volcano is a low, rounded mountain formed when lava spreads out around a vent in a quiet eruption.

shoreline, page 56.

The shoreline is the place where the land and the ocean meet.

silt, page 90.

Silt is soil made when quartz is broken down. Silt has tiny grains.

slope, pages 56, 62.

The slope of the ocean floor is a steep hill from the shelf to the abyssal plain. A mountain slope is a side that curves down.

smog, page 130.

Smog is a kind of hazy air pollution that often hangs over cities.

soil, page 90.

Soil is made of broken bits of rock mixed with air, water, and humus.

solar system, page 4.

The sun and the nine planets make up the solar system.

source, page 46.

The place where a river starts is called its source.

South America, page 60.

South America is one of the seven continents. It is connected to the continent of North America.

South Pole, page 8.

The South Pole is the southernmost point on Earth.

spacecraft, page 18.

A spacecraft is a vehicle that can travel in space.

stalactite, page 52.

A stalactite is a long pointed rock that hangs from the roof of a cave.

stalagmite, page 52.

A stalagmite is a long pointed rock that rises from the floor of a cave.

star, page 6.

A star is a huge ball of hot gases. It is made up mostly of hydrogen and helium.

strata, page 114.

Strata are layers of rock.

stratosphere, page 26.

The stratosphere is the layer of the atmosphere that reaches from the troposphere to 30 miles above Earth.

stratus, page 30.

Stratus clouds are layers of gray clouds that cover most of the sky.

strip cropping, page 128.

Strip cropping is a way farmers prevent soil erosion. Rows of different crops are planted across a field.

subsoil, page 90.
Subsoil is a layer of soil made of larger rocks that have been broken down.

 telescope, page 18.
A telescope is a tool that makes distant objects look larger.

temperate, page 68.
In a temperate climate zone, winters are cold and summers are warm.

temperature, page 35.
Temperature is a measure of how hot or cold something is.

thermosphere, page 26.
The thermosphere is a layer of the atmosphere where the air becomes thin and fades into space.

thunder, page 40.
Thunder is the noise that air makes when it is heated by lightning and expands quickly.

thunderstorm, page 40.
A thunderstorm is a storm with heavy rain, thunder, and lightning.

tide, page 14.
A tide is a change in the level of water in the oceans.

topsoil, page 90.
Topsoil is the uppermost layer of soil. It has rock particles, humus, and minerals.

tornado, page 40.
A tornado is a violent but small storm that starts over land.

trench, page 98.
A trench is a deep valley formed when two plates under the ocean floor collide.

tributary, page 46.
A tributary is a river that flows into another river.

tropical, page 68.
In a tropical climate zone, it is warm all year round and rainfall is high.

tropical rain forest, page 72.
A tropical rain forest is a forest that grows where it is warm and wet all year.

troposphere, page 26.
The troposphere is the layer of the atmosphere closest to Earth. Nearly all weather occurs in this layer.

tsunami, page 102.
A tsunami is a giant wave produced when the focus of an earthquake is under the ocean floor.

tundra, page 78.
Tundra is a huge area of plains with no trees.

U-shaped valley, page 86.
A U-shaped valley is formed as a glacier moves.

V-shaped canyon, page 84.
A V-shaped canyon is formed when flowing water wears away a riverbed.

valley, page 62.
A valley is a low place between hills or mountains.

vent, page 104.
A vent is an opening in a volcano.

volcanic mountain, page 100.
A volcanic mountain is formed when layers of lava pile up on the surface and cool.

volcano, page 104.
A volcano occurs when material from the mantle erupts onto the surface.

water cycle, page 28.
The water cycle is the movement of water between the ground and the atmosphere.

water table, page 50.
The water table is the top of an underground stream or reservoir.

water vapor, page 28.
Water vapor is water in the form of a gas.

wave, page 84.
A wave is the motion of ocean water.

weather, page 26.
Weather is the condition of the air around us.

weathering, page 82.
Weathering is the breaking up of rocks and soil on Earth's surface.

well, page 50.
A well is a hole dug deep enough to reach the water table. Water is pumped up through the well.

wildlife refuge, page 134.
A wildlife refuge is a place where plants and animals are kept safe.

wind, page 38.
Wind is the movement of air from a high-pressure area to a low-pressure area.

year, page 20.
A year on Earth is 365 days. It is the time it takes Earth to orbit once around the sun.

WONDERS OF SCIENCE

Acknowledgments

Staff Credits

Executive Editor:	Stephanie Muller
Senior Editor:	Jayne Cotton
Design Manager:	Richard Balsam
Designer:	Jim Cauthron
Photo Editor:	Margie Foster

Illustrations

Erika Kors—**50, 54, 56, 58, 60, 62, 64, 68, 90, 96, 98**

Tek-Nek, Inc.—**38T, 38B**

Stephen Turner—**8, 20, 21, 26, 28, 36, 77**

David Griffin—**24, 44, 66, 80, 94, 108, 124, 136**

All Other Illustrations

Joe Nerlinger and Lewis Calver

Photographs

Cover and Title Page: © West Light/Charles Campbell

P.**6** NASA; p.**10 (a & b)** Jet Propulsion Lab, **(c & d)** NASA; p.**14 (all)** Lick Observatory; pp.**16, (both), 18** NASA; p.**30 (a & b)** National Oceanic and Atmospheric Association, **(c)** © Grant Heilman Photography, **(d)** © David Dennis/Tom Stack & Associates; p.**32 (a)** © Margaret Brandow/Tom Stack & Associates, **(b)** National Oceanic and Atmospheric Association, **(c)** National Center for Atmospheric Research/National Science Foundation; p.**34 (a)** © Thomas Mangelson/Images of Nature, **(b)** © P. Mason/Instock; p.**40** National Center for Atmospheric Research/National Science Foundation; p.**46 (a)** © Alan Pitcairn/Grant Heilman Photography, **(b)** NASA/Grant Heilman Photography; p.**48** Bureau of Reclamation, US Department of the Interior; p.**52** © Brian Parker/Tom Stack & Associates; p.**70** Texas Highways; p.**72 (a)** Conservation Commission of Missouri, **(b)** © Stephanie S. Ferguson/William E. Ferguson; p.**76** © Grant Heilman Photography; p.**78** © Manuel Rodriguez; p.**82** Utah Travel Council; **p.84 (both)** US Geological Survey; p.**86 (a)** US Geological Survey, **(b)** Geological Survey of Canada; p.**88 (left)** National Park Service, **(right)** Reuters/Bettmann; p.**92** National Coal Association; p.**100 (a)** Utah Travel Council, **(b)** Aspen Skiing Corporation, **(c)** Utah Geological and Mineral Survey; p.**102** UPI/Bettmann; p.**104** © Mark Hurd Aerial Surveys; p.**106** National Park Service; p.**110 (all)** © Grant Heilman Photography; pp.**112 (all), 114** US Geological Survey; p.**116 (a)** Ward's Natural Science Survey, **(b)** US Geological Survey; p.**118** © Manuel Rodriguez; p.**119 (a)** © Jim Coxe/American Museum of Natural History, **(b)** © Tom Bean/DRK Photo; p.**120 (a)** Monterey Coal Company, **(b)** © Grant Heilman Photography; p.**122 (a)** Monterey Oil Company, **(b)** Texas Highways; p.**126 (a)** © Grant Heilman Photography, **(b)** Texaco, Inc.; p.**128** © Grant Heilman Photography; p.**130** © Eda Rogers/PhotoEdit; p.**132 (a)** © JG/PhotoEdit, **(b)** US Fish and Wildlife Service; p.**134** Texas Highways.